Thirty-Eight

Thirty-Eight

The Hurricane That Transformed New England

Stephen Long

Yale UNIVERSITY PRESS

New Haven & London

Published with assistance from the foundation established in memory of Calvin Chapin of the Class of 1788, Yale College.

Yale University Press books may be purchased in quantity for educational, business, or promotional use. For information, please e-mail sales.press@yale.edu (U.S. office) or sales@yaleup.co.uk (U.K. office).

Set in Janson Roman type by Integrated Publishing Solutions, Grand Rapids, Michigan.
Printed in the United States of America.

Library of Congress Control Number: 2015945291
ISBN 978-0-300-20951-8 (cloth : alk. paper)

A catalogue record for this book is available from the British Library.

This paper meets the requirements of ANSI/NISO Z39.48-1992 (Permanence of Paper).

10 9 8 7 6 5 4 3 2 1

To my father, who gave us summers in the woods,
and to my mother, who made those years so much fun
Thomas Long (1911–2001)
Mary Beers Long (1916–2013)

Contents

Contents

Preface

The hurricane that pummeled the East Coast on September 21, 1938, was New England's most damaging weather event ever. Call it New England's Katrina and you might be understating its power. The storm plowed into Long Island and New England without warning, killing hundreds of people and destroying roads, bridges, dams, and buildings. The devastation to the region's infrastructure required repairs costing $300 million in Depression-era dollars, approximately $5 billion today.

Almost every word that has been written about the 1938 hurricane recounts the damage to the built environment and to the people who lived in it. It has been an urban story rather than a rural one, a tale of the coast rather than the inland forest. That's understandable. Compared with the loss of human life and the destruction of property, damage to trees might seem like a scratch on the fender of a car that's been totaled. Still, our lives depend—either directly or indirectly—on forests, and the destruction of a thousand square miles of forestland remains a story that needs to be told.

Forest ecologists have long seen Thirty-Eight as a touchstone event and have studied its long-lasting impacts across the region, but their impressive work on the role of huge disturbances like this has not reached the layman. Little has been written for public consumption about what happened to the forests of New England, and some of it has been so casually reported that it's grossly inaccurate. In a book titled *Hurricane!*, Joe McCarthy (no, not that Joe McCarthy) notes that New Hampshire lost half of its white pines, which is a reasonable statement. But McCarthy follows that with this whopper: "Most of the timber felled by the hurricane was too splintered to be used, except as firewood or as pulp at a paper mill." In fact, the hurricane recovery effort awakened a moribund wood-products industry from the Depression. The administrators of Franklin Roosevelt's New Deal were accustomed to big challenges and had shown no timidity in putting the full force of the government into play, so the U.S. Forest Service jumped in and created a new agency to respond to the emergency. It expedited the salvage of billions of board feet of logs, but it had another reason to rid the area of the limbs, branches, and boles that carpeted New England. Forest Service officials feared a wildfire that, once ignited, could spread throughout the region, killing more people and destroying more infrastructure. The scale of the timber salvage and fire hazard reduction meant that the region's worst hurricane was followed by the largest logging job the country has ever seen.

In *Thirty-Eight* I tell the story of this unprecedented one-two punch through a number of lenses: forest ecology, meteorology, social science, political science, and land management. An event of this magnitude requires a multifaceted narrative because the natural disaster and the human history—before and after—cannot be separated. Ecology and economy have never been wound more tightly together.

The last time a storm of anywhere near this magnitude had hit

New England was more than a century before, in 1815. So many generations had come and gone since then that the possibility of a hurricane had disappeared from the public consciousness, and newspapers the following day proclaimed it New England's first hurricane. Such an extended period without hurricanes seems inconceivable, given our recent history of extreme weather events. Meteorologists and forest historians tell us that another hurricane of this magnitude could occur at any time. We live in an entirely different world from the one laid low in 1938. In *Thirty-Eight* I examine the ways those differences will influence how destructive the next hurricane will be.

If you have an interest in New England history, forests and trees, or extreme weather events in a changing world, pull up a chair. This is a book about forests and people. We are part of nature. We have changed nature profoundly, but still our forests sustain us economically, ecologically, and spiritually. In *Thirty-Eight*, I tell the story of a remarkable hurricane and show how a combination of natural disturbance and human actions has created the world we live in. More than seventeen million people now live within the area affected by the hurricane. The green landscape that surrounds us might seem like a comfortingly stable backdrop for daily life, but it is a dynamic system changing every day. And every once in a great while, an explosion of change bursts across a wide area in a way that people never forget.

Acknowledgments

I completed much of the research for this book as a Charles Bullard Fellow at Harvard Forest. I am grateful to the researchers and staff at that very welcoming institution, who were so helpful and a great pleasure to work with. Special thanks to Dave Orwig, Emery Boose, Audrey Barker Plotkin, Elaine Doughty, Julie Pallant, John O'Keefe, Clarisse Hart, Scot Wiinikka, Edythe Ellin, and Laurie Chiasson. Brian Hall stepped up with his considerable mapping skills and created the illustrations that help explain the meteorology. David Foster, Harvard Forest's director, deserves superlatives in many regards: inspiring teacher and leader, prolific writer, and highly influential figure in the world of ecology and conservation, he still always found time to help me understand the nuances of natural and human disturbance. He has opened many doors for me, and for that, I am indeed grateful.

I thank Bob Saul, whose encouragement, enthusiasm for this project, and generosity made it possible for me to set aside the time I needed to complete the writing.

Acknowledgments

I am indebted to the many people who told me their stories of living through the Great New England Hurricane, especially Harry and Joan Brainerd, Dustin and Jane White, Fred Hunt, Bryce Metcalf, Lois Sherwood, Jim Colby, Allen Britton, Tink and Polly Hood, Fred Spooner, Harold Luce, John Hemenway, and Put Blodgett. My gratitude to Fred Hunt, Jim Colby, Harry Brainerd, and Harold Luce is tempered with sadness; they were instrumental in helping me understand the terror of the storm and have passed away before they could hold this book in their hands.

Some other people you will meet in the book have been enormously generous with their time, particularly Charlie Cogbill and Lourdes Aviles, who put up with an endless stream of follow-up questions, and local weatherman Mark Breen, who gave me a crash course in meteorology.

Much of the information on the salvage program came from the National Archive in Waltham, Massachusetts, a repository of hundreds of boxes of memos and letters of the Northeastern Timber Salvage Administration. The archives at Harvard Forest hold many important records, including before-and-after inventories of the forest's various compartments. Two online archives have also been invaluable: the National Oceanic and Atmospheric Administration (NOAA) for all things meteorological and the Forest History Society for forestry and the New Deal.

Many individuals, including staff at libraries, historical societies, and other organizations, have helped me find documents and articles for my research and photos for use in the book. Though too many to name, you all have my gratitude.

Ginny Barlow, Dave Paganelli, Mike Batten, and Markus Bradley have helped me manage my woodland and taught me so much about what to look for and how things work in the woods. Everything I write springs from that base of knowledge. Thanks also to my former colleagues at *Northern Woodlands* magazine, which published an early story based on some of this research. Sarah Smith generously shared what she learned researching her own book, *They Sawed Up a Storm*. Writer and naturalist Charles Johnson has long championed my work, and I want to thank him publicly for his warm encouragement.

My wife, Mary Hays, disrupted her own writing life to join me in my fellowship year at Harvard Forest, traveling home and back with me each week while our restive dog panted in the back seat. She always is my closest and best reader. I've benefited from other readers as well—David Foster, Charlie Cogbill, Lourdes Aviles, Bob Saul, and Andy Morris—who have pointed out my exaggerations and misapprehensions. Finally, many thanks to some key people at Yale University Press: Jean Thomson Black for her enthusiasm for this book and her guidance along the way; Samantha Ostrowski for her steady efficiency; and Dan Heaton for his insightful and subtle editing. All of these people have helped make this a better book. I take sole responsibility for any deficiencies that remain.

Thirty-Eight

ONE

White Pine

In the photo, Fred Hunt stands in front of a stack of perfect sawlogs. He's all puffed up with four layers of shirts and jackets, but no hat or gloves. A lifelong outdoorsman in his eighties, he has only recently begun to be bothered by the cold, and he is trying to stay warm on a cold but bright February day. He'd been showing me and a colleague some of the impressive white pine he was growing at Sylvan Acres, his 750-acre tree farm in Reading, Vermont. We'd seen acres of large-diameter trees reaching 120 feet into the sky, not a branch on them for the first 40 feet. Leaving the woods, we came to the log landing and the pile of logs where I shot the photo. Hunt's son, Bill, had cut these massive trees earlier that week and stacked them on the landing. Hunt explained that the logs filled a special order for a log-house builder named John Nininger, who specializes in buildings made of large logs.

Hunt said, "I'm not talking log cabins, I'm talking about big houses for millionaires trying to outdo the other one with who can have the biggest log house." White pine is the only tree in the Northeast that routinely grows to this size. Even so, perfectly straight specimens like these logs—twenty feet long with at least a twenty-four-inch diameter at the small end—are anything but routine, and Nininger knows he can find what he wants at Sylvan Acres. Hunt sends him off into the woods with a paint can to select his trees, and then Bill Hunt cuts and skids them to the landing. Over the years, Hunt's logs have gone into dozens of Nininger's houses.

The old forester is squinting and not looking at the camera in any of the shots, so you see a three-quarters profile. With his craggy eyebrows and his granite jaw still absent of jowls, he's a true son of New Hampshire, though it turns out that despite his many years in the Granite State, he was in fact born across the line in Massachusetts. He's tucked in tight to this pile of logs, and man and logs are brightly lit by the midmorning sun. One benefit of the brightness is that the deep focal length of the lens has kept both Hunt and the logs in tight focus. So tight, in fact, that you can count the rings on the end of the log over his right shoulder. Which is how I know that it came from a tree planted before Fred Hunt was born.

This particular pine is twenty-seven inches in diameter, and counting inward from the bark you can see a change in the growth rings the year after Hunt bought the land in 1955. The fifteen years before his purchase had been poor growing years for the pine. The rings grew close together, showing that it took five years to grow an inch. Then there's a remarkable increase in 1956, and the growth rings from then on are almost uniformly wide. It's as if he willed this tree to grow faster. In a way, that's exactly what he did, if by willing it to grow you mean working six days a week constantly weeding out the crooked, forked, poorer-quality trees. Under Hunt's nurturing, this particular tree put on nineteen inches of growth in fifty years, just about four inches per decade.

"I'm a white pine man," Hunt told me, but it took a lot of probing —which proved a challenge with this notoriously tight-lipped man

Fred Hunt at the age of eighty-two with white pine logs cut from his Sylvan Acres.
He was fourteen when the 1938 hurricane blew down around him in Rindge,
New Hampshire. Stephen Long

—to learn the extent to which that was true. After graduating with a degree in forestry from the University of New Hampshire, he ran a logging business for ten years, specializing in thinning pine plantations. He then earned an M.S. and Ph.D. from the University of Massachusetts, studying white pine and its effect on the water supply. Along the way, his master's thesis served as the first management plan for the 58,000-acre forest surrounding Quabbin Reservoir, which provides the drinking water for Boston and forty nearby towns. He then taught forest management and managed the 12,000-acre forest at Paul Smith's College deep in the Adirondacks for ten years before he decided, at the age of fifty-four, to make his final career change. His rationale: "I had 750 acres of my own land in Reading. I couldn't afford to spend the rest of my life telling everybody else what they should do and let my own woods go to pot, so I came back to Reading and basically have worked there ever since."

Hunt spent a lifetime working to grow superior white pine because it provided a good living and because he loved the practice of silviculture. But it's possible that his lifelong affinity for white pine could have little to do with money or forestry. It could have more to do with an event when he was fourteen years old. On that day, a white pine saved his life.

By the morning of September 21, 1938, it had been raining for days. Not exactly the kind of day most kids would choose for playing hooky, even if you loved the outdoors as much as Fred Hunt did. You couldn't fish, the rivers were too swollen, but that's just what drew him to go on a lark. He'd heard that Jaffrey, the big town five miles to the northwest of his home in Rindge, was under water. Never having seen a flooded village, he pulled on his rubber boots and left his home on Old New Ipswich Road to go inspect the flood damage. "I was basically homeschooled, so if I took off, there wasn't any school that day," he told me.

Hunt was recounting this early example of his lifelong contrariness while sitting in the living room of his house in Reading. It was now September 2011, not quite five years since I'd taken his picture after the tour of his woods. He was now eighty-seven, and age had caught up with him. Three days a week, his daughter drove him to dialysis appointments at Dartmouth-Hitchcock Medical Center, a forty-five-minute drive each way. His voice was faint, but he was willing to talk. He was even willing to continue our tour. He had more to show me, places we hadn't gotten to on my previous visit.

But I demurred, and not only because of the steady rain we could

see dripping from the eaves. I was there to hear about his day playing hooky. And even though it had occurred seventy-three years
earlier, he could recall his route mile by mile, incident by incident.
I'd brought along a photocopy of the Cheshire County page from
the *New Hampshire Gazetteer* to spark Fred's memory, but he needed
no spark. The hike he described might as well have happened the
day before.

Rindge and its part of Cheshire County are dotted with ponds
and lakes, large and small. Besides Hubbard Pond, a beautiful lake
virtually in the Hunt family's backyard, there are Crowcroft Pond,
Grassy Pond, Emerson Pond, Bullet Pond, and Pool Pond, not to
mention Contoocook Lake and Black Reservoir. The hills between
them are not particularly tall except for the singular Mount Monad-
nock, which rises to twenty-eight hundred feet northwest of Jaffrey.
Hunt's childhood home on Old New Ipswich Road stood at twelve
hundred feet, while Jaffrey sits in a valley two hundred feet lower.
Depending on which part of town you hiked through, the falling
rain could end up either in the Atlantic Ocean or in Long Island
Sound. One watershed sent the water north and east via the Contoocook and the Merrimack to the ocean; the other drained west
and south via the Millers and the Connecticut to the sound.

The rain swelled the ponds and streams that morning. Starting
out at about 7:00 A.M., he trekked around the west side of Hubbard
Pond, but at the head of the pond, the brook that drained it was
already over the road by then. So he turned around and followed
an old logging road down through to the village of Squantum, still
wanting to get to Jaffrey. But he was stymied again by high water.

It was near Squantum that he helped to rescue a pig. "This farmer
had a pig yard out back of a sawmill. We roped him up and dragged
him across through the sawmill and out onto the high land on that
side. And one of the fellas, after we got the pig rescued, tried to go
back across where we just dragged the pig from. He couldn't cross

even with a rope tied around his waist and people holding the rope. He got partway and he wouldn't try it."

Hunt's next idea of going up over the hill and avoiding all the lowland didn't pan out either. At some point he had to drop down, and every place he did was flooded. The water at Route 202 stood higher than the highway fence posts. So he abandoned his quest for Jaffrey and settled on Winchendon, in the other direction and just down over the Massachusetts border. He took off down the old Route 202 (the current 202 has been built since then a few miles to the west) and made it to the state line. There he found the highway buckled and a whole section of it with the yellow center line pointing to the top of the maple tree it was leaning on. The western overflow outlet of Lake Monomonac, a six hundred–acre lake straddling the state line, had been breached. "All that water went down into the Millers River," Hunt said. He was impressed.

That's when he started back home. His socks were soaked inside his rubber boots, and his heels were blistered raw. By the time he got to Rindge Center, he'd had enough, so he stopped in a store there and telephoned his mother to come pick him up. He calculates that he had hiked somewhere between fifteen and twenty miles that day.

By the time his mother reached the store late in the afternoon, the rain had returned and the wind was howling. As they drove home, their car was blocked by a small tree across the road. So Fred and his mother stopped at the neighbors' house, which sat up high on the hill, exposed to the wind. They found the three women who lived there trying to contain a stream of water pouring through the attic. All the shingles had blown off the roof, and the water was gushing in. Fred's mother stayed there with the neighbors, but she sent him home on foot the remaining half mile because his father was there alone.

"I decided that stuff was flying too fast up there on the hill so I

Thousands of stands of white pine looked like this when the sun shone the following day. This stand was in Middlesex County, Massachusetts.
United States Forest Service

went down across the hayfield and down into the woods and back to the town road at the bottom of the hill," he said.

The fourteen-year-old boy was nearing home, walking down a road that was lined, like so many in New England, by stone walls on both sides. The white pines on either side were fifty to seventy years old and stood tall. But not for long.

"And that's when I happened to look behind me and there was this pine tree about yay big" (his arms spread wide like a fisherman demonstrating the size of the one that got away) "across the road about ten or fifteen feet behind me. The diameter was two feet anyway, and maybe three. I don't know, I'm looking back now seventy-odd years, but it was a good-sized tree, and it just missed me, because I wasn't fifteen feet beyond it when it came down."

I asked him how such a huge tree could fall without him hearing it. He told me to try to imagine wind blowing so loudly and steadily that you wouldn't notice a three-ton tree crashing down five steps away. "With the roar from that wind you couldn't hear anything else, it was just a roar." In an instant, he realized that the huge tree straddling the road with its top perched on a stone wall could be his refuge from whatever else might happen. So he scrambled underneath it.

"I was under that pine tree when the hurricane came through," Hunt said. I can picture the situation clearly as he tells the story, and I admire his quick thinking because I'm not sure I would have recognized that fallen tree as a safe haven. He spent ten minutes, maybe fifteen, beneath that tree while every other tree in the forest came crashing down. Eventually, he crawled out from under it and poked his head up. There wasn't a single tree standing. "My estimate was that 90 percent of all the trees in Rindge over six inches in diameter got blown down in ten minutes or so. Some of them broke off, but most of them uprooted," he said. "Most of it was pine."

Experience has shown me repeatedly that the notion of the taciturn Yankee is a fiction. They're not all descendants of Calvin Coolidge. Ask almost any elder a question, even with a tape recorder running and a microphone in plain sight, and once they get started, you'll have a hard time sneaking in another question. They have so much to say, and they're happy to have an interested listener. Fred Hunt, however, as in nearly every regard, ran counter to the norm. Friends in later years suggested that his mother's homeschooling had encouraged his inquiring mind but had done nothing to develop the social skills he would have learned in a classroom with his peers. Try as I might, I couldn't get much more out of him. He had nothing more to say about what it felt like to be under that tree while New Hampshire's worst hurricane blew down every single tree around him. I asked him the obvious question: Weren't you scared?

"People always ask, 'Were you scared?' No I wasn't scared. There was nothing to be scared of. That first tree missed me. Then it all came down with a rush and a roar. Another five or ten minutes and there wasn't anything left to blow down. It all went *whooossh*."

The wind continued to roar, but the damage had been done. "Once I could see there wasn't much of anything to blow down, why, then I crawled my way through it for the rest of the way home. I went home and stayed with my father."

One point that Hunt took pains to make clear through his story of checking out the flood damage is something that has often been obscured in accounts of the 1938 hurricane: much of New England was already flooded by an entirely different weather system the day the hurricane hit. Early that day, it showered off and on, but the rain was nothing like it had been the night before. In fact, the water was receding, as he could tell from the high-water marks he saw everywhere he went. If the hurricane and all of its wind hadn't arrived later that day, the September weather event would have been known as the flood of 1938.

The roaring wind toppled forests in every New England state, with New Hampshire and Massachusetts hit particularly hard. The path of destruction spanned ninety miles across, its western boundary starting near the Connecticut River Valley in Connecticut, drifting slightly westward and then following the spine of the Green Mountains into Vermont. Damage to the west of this line was largely from rain rather than wind.

All told, an estimated 2.6 billion board feet of timber was blown down. I'm sure that sounds like a lot, but numbers of that magnitude require some context to be comprehended. Board feet is a cubic measurement, and one board foot is a one-inch thick board twelve inches long and twelve inches wide. You've probably handled a two-by-four; an eight-foot-long two-by-four totals 5.3 board feet. If it were 16 feet long, it would be 10.6 board feet. Logs are also measured in board feet, and a typical logging truck today carries 6,000

This log truck, typical of those operating throughout New England today, carries a load of six thousand board feet. The hurricane blew down the equivalent of 430,000 truckloads of timber. Stephen Long

board feet of logs. It would take 430,000 of these trucks to transport the wood that was blown down that day.

Here's another way to think of it. The largest white pine sawmill operating today in New England is Pleasant River Lumber in Dover-Foxcroft, Maine. It buys 90 million board feet of spruce and pine sawlogs annually. Each week on average, 288 log trucks arrive at the mill. At their current rate of production, it would take Pleasant River twenty-nine years to process 2.6 billion board feet.

One last yardstick, and the one that has the most relevance: the combined annual harvest in the four states whose forests were hit hardest by the hurricane—Connecticut, Massachusetts, New Hampshire, and Vermont—totaled 500 million board feet in the late 1930s, so the storm laid down five years' worth of logs. As the industry crawled out of the slump that began in 2008, the harvest in those four states has been around 1.2 billion board feet. So the annual harvest in a business-as-usual fashion using today's highly productive equipment—feller-bunchers and grapple skidders, or chainsaws and cable skidders—is less than half of what went down in a five-hour period on September 21, 1938.

■

It's hard to imagine that one species could be singled out so completely for destruction, but 70 percent or more of the toppled timber was *Pinus strobus*, eastern white pine. Connecticut and Vermont were the only states where hardwood species like maple, oak, beech, and birch made up a considerable portion of the loss.

In pondering why so much white pine blew down, I'm reminded of the exchange in *The Wild One* between Marlon Brando's character, Johnny, and one of the local girls. She asks innocently: "What are you rebelling against, Johnny?" Brando looks back, "Whaddya got?" If in 1938 you asked a forest owner in Massachusetts or New Hampshire, "Whaddya got?" the answer would have been white pine.

We don't have reliable inventory information that would tell us the precise makeup of New England's forest in the late 1930s, but we know that Massachusetts and New Hampshire had more pine at this time than they had ever had. One important cause of this was the way the land had been used over the generations since Europeans arrived in the seventeenth century. The forest that greeted them covered 95 percent of the landscape. Settlers found this bounty of timber useful for building houses and barns, for fencing in animals, and for the twenty cords of fuel they needed to heat their drafty houses. There was a more compelling reason to carve up the wilderness: it was in the way. The continued livelihoods of these subsistence farmers depended on clearing land to grow plants and animals. Opening up a patch of woods for a farm was repeated generation after generation, as grown children struck out on their own, pushing farther and farther into the interior of New England. With hundreds of thousands of pioneers clearing their own patches of forest with axe, saw, and fire, it took more than two centuries to open up most of Connecticut, Rhode Island, Massachusetts, and accessible parts of New Hampshire, Maine, and Vermont.

They farmed every place that they could, using each acre according to its workability. The best soils on the flattest ground were plowed to plant corn, wheat, and oats. Land that resisted the plow but was clearly productive could be sown with grass and clover and cut twice or more a year for hay. Steeper, wetter, rockier land served as pasture for animals grown for meat or for work. All of the land within this hierarchy of use—cropland, hayfield, pasture—was cleared of trees. In addition, most farmers had some land in their fifty- or one hundred–acre piece that was so wet, steep, or rocky that it wasn't worth clearing, and they kept these as woodlots to provide an ongoing supply of hardwood fuel or softwood building materials.

The most thorough clearing occurred where people had lived the longest, along tidal rivers. Settlement was a rising tide that flowed up the rivers from the coast deeper into the interior with each generation. The sons and grandsons of Massachusetts pushed into New Hampshire, while their counterparts from Connecticut founded many of the towns in Vermont.

Sawmills sprang up along rivers in most communities, followed by other commercial enterprises. As frontier towns became fully settled, societal and economic development meant that the population was no longer exclusively farmers. By the nineteenth century, southern New England had become more industrial, as millers harnessed water power to run mills of all kinds. Formerly subsistence farms grew surplus meat and grain that were shipped to cities. Demand for timber meant that forests continued to be cut hard, even from relatively inaccessible slopes and hollows. Sons and daughters who weren't to inherit or marry into a farm moved for jobs in the increasingly industrialized larger towns and cities. Others moved out of New England altogether, part of the wave that settled in the territories west of the Appalachians and beyond.

The land was being put to its most intense use in the decades either side of the Civil War. Half or more of southern New England was cleared for agriculture, with livestock pastured on most of the farmland. Woodlots were cut hard and repeatedly for fuel, so most of the remaining forest was young. Farmers even pastured cattle in woodlots to browse on hardwood saplings. This intensity of use couldn't last forever. The peak of cleared land in Massachusetts came around 1850, and a wholesale shift away from using the land for farms was well under way by 1880. A number of factors have been cited to explain how agriculture's supremacy began to fade in New England. Mechanization of farming produced more yield per acre especially in the deep soils of the Midwest. Despite centuries of Yankee farmers using fieldstone for walls, foundations, and other projects, the ground still produced a crop of rocks each year. Hilly, rocky New England couldn't compete with the better soils out west. Meanwhile, the Erie Canal and then the railroads made it possible to ship commodities grown in western states to compete with those grown locally.

It didn't happen overnight, but the agricultural use of the land subsided, with the marginal farmland going out of production first. The phenomenon is known as farmland abandonment, but this term needs some explanation. It's not that farmers walked away from their land. Even while much of the land was still being farmed productively, the poorer land—cleared by the earliest settlers in a fit of exuberance or necessity—proved itself not to be worth farming. If the current owner could get some cash for fallow land from a neighbor who saw more potential in it than he did, they made a deal. Abandonment occurred at the margins, and there was always a buyer willing to speculate on what the seller perceived to be worthless. Prosperous farmers consolidated their neighbors' lands into their own, and populations shrank in rural towns.

Most of this land reverted to forest. Almost everywhere in the Northeast, the default vegetation cover is trees. If you don't prevent them from growing, they return. Stop mowing, stop pasturing sheep or cattle, stop cutting small coppice wood for cooking stoves, and the forest will return. In this way, forests regained control of a landscape that had been painstakingly cleared a few generations before. Predictably, the trees that seeded into these old fields were those known as pioneer species. These include paper birch, several aspens and several cherries, and white pine. These species produce prodigious volumes of seed, and though they prefer a seedbed of disturbed mineral soil, they can and will colonize a field covered with grasses and forbs. More than other tree species, these early successional species thrive in the ample sunlight of a former pasture. Birch, aspen, and cherry are the champions at taking advantage of their moment in the sun, and they can put on impressive height growth right away. White pine is a bit of a second-stringer in this regard, and it might take pine seedlings a decade to reach a height of five feet. But pine has an important advantage over the hardwood species. If the cows or sheep were turned out into those reverting pastures for even a short time, they would browse the hardwoods and leave the less-palatable pine alone. Pine can also thrive in dry conditions and compete effectively with grasses. Once it establishes itself in an opening with abundant sunlight, it can outpace any other tree in the Northeast's forest. And if it escapes the saw, pine can outlive its pioneering compatriots. White birch lasts seventy or eighty years before falling apart, and aspen doesn't make it even that long. White pine can routinely live two hundred years—unless, of course, it's blown down by a hurricane.

In most cases, these sun-loving pioneer species are not the same species that were cut when the forest was cleared a few generations back. White pine was an important target species for its many commercial uses, but it was never a large component of the forest. Charlie Cogbill, an expert on the presettlement forest, told me, "There were areas along the rivers that had very useful, merchantable pine. Otherwise, in the uplands they didn't have very much or had a scattered tree here or there." Mythmaking has doubtless played a part in the popular notion of the prominence of white pine. For instance, there is the tradition of the mast trees marked by the King's Broad Arrow. No matter who owned the land, huge pines were reserved for the king of England for ships' masts for the royal navy. They were designated by a broad arrow emblazoned on the trunk of the tree. The dictum was unevenly administered, and in one case where the British officials clamped down, it inspired a revolt among colonists in Weare, New Hampshire, in 1772. Doubtless, there were grand pines that towered over the rest of the forest, and accounts of their magnitude may have given an impression of plenitude. Almost nowhere was it a forest full of pine. Cogbill's research shows that even in Maine, known as the Pine Tree State, it accounted for less than 5 percent of the trees at the time of settlement. Across New England, white pine may have constituted 7 percent of the total.

Across much of the land, the original forest was a mature forest with a mix of trees dominated by those most tolerant of shade. Shade tolerance is one of those forestry terms that mean exactly what they say. It's not that these species—in particular, sugar maple, eastern hemlock, and American beech—prefer shade. They don't. All trees want to gain as much sunlight as they can. But these species *tolerate* shade, giving them a long-term advantage over the pioneer species, which cannot reproduce in their own shade. For the intolerant species, it's one and done. They thrive for a single generation in any one place, and their pioneering work creates site conditions that favor the more tolerant species that take over from them. This is forest succession at its simplest: an intolerant species takes over a site and is succeeded over time by a species with more shade tolerance. White pine has a unique niche. Though it functions as a pioneer species, it has intermediate shade tolerance: it tolerates shade better than birch and aspen but not as well as maple and beech.

A shade-tolerant species can control its turf for many generations. Beneath a full canopy that intercepts all but transient beams of light, maple or hemlock seeds will germinate, seedlings will grow into saplings, and saplings into poles and larger. Their leaves have the capacity to photosynthesize at very low light levels, and suppressed trees can persist in the understory for decades, waiting for the death of a large neighbor that will allow them to reach a spot in the canopy. A climax forest is one that maintains a stable mix of species for several generations. By definition, it is dominated by long-lived shade-tolerant species that can succeed themselves in the diminished light of a full canopy. Pioneers can gain a foothold when small holes in the canopy open up periodically from local microbursts and other natural disturbances, but a climax forest can develop only if it remains free of any widespread disturbance—fire or a hurricane—for three hundred or four hundred years. Otherwise, those disturbances would open up the canopy and provide an opening for the pioneers, restarting the process of forest succession.

The Europeans' clearing of land for farming dwarfs hurricanes and fire as a profound disturbance. It irrevocably altered the entire natural system. European plant species, for instance, came over in the hay that accompanied the shipments of cattle. Walk in almost any hayfield or pasture today and you'll witness their progeny: dandelions, white clover, Queen Anne's lace, mullein. These nonnative

White pine has found this abandoned pasture a hospitable place to seed into.
Most of the pine that blew down in Thirty-Eight was so-called old-field white
pine, which had established itself thirty to fifty years before the hurricane struck.
Stephen Long

species are relatively harmless plants, to be sure, except that they
replaced not only native grasses but trees. Dandelions and clover
surrounded the stumps of oaks and maples, and soon took control.

Similarly, populations of many woodland animal species plum-
meted from the loss of forest and the hunting that accompanied it.
Moose, deer, turkeys, wolves, and mountain lions disappeared, and
deep woods songbirds like the hermit thrush and scarlet tanager were
much scarcer than before settlement. Many New Englanders never
heard the dawn and dusk serenade of the wood thrush. Replacing

them were grassland birds like bobolinks and meadowlarks, and mammals that thrive on agriculture: foxes, skunks, and raccoons. Depending on its date of settlement, in most places the new regime lasted for at least a century before the pendulum swung back.

The year that a pasture is abandoned, the forest starts to grow back. Trees happen. Abandoned fields—five acres here and twenty acres there—provided the perfect conditions for white pine to develop into almost pure stands as they outpaced any of the hardwoods that seeded in at the same time. The regeneration of Massachusetts and adjacent New Hampshire with thousands of acres of old-field white pine began in the middle of the nineteenth century after the agricultural use crested. As the decades rolled by, old fields continued to be abandoned. Fields of pine sprang up every year in every town, and by the dawn of the twentieth century, the proportion of white pine in the canopy reached all-time highs.

■

The tree that saved Fred Hunt began his lifelong love affair with white pine. At fourteen, the adventurous young man already had an entrepreneurial bent, and he wasn't afraid of hard work. After wading through the tangle of limbs and branches on the property next door, he struck a deal with the owner. He took on the job of cleaning up all the pine brush, and in return he could do anything he wanted with the logs. "I salvaged all the trees, hired someone to skid the logs out, hired somebody to truck logs to the mill, and hired a mill to saw the lumber out," said Hunt. This was all accomplished before his fifteenth birthday.

Despite his professed lack of enthusiasm for the classroom, he graduated from high school. He was deferred from military service during World War II because his draft board physical turned up an enlarged heart, a condition that hadn't bothered him before and never did. He fully expected to serve, but when he was rejected, he enrolled in the forestry program at the University of New Hampshire, one of only four who entered that year. Already one to trust his own experience over what had been handed down by others, he laughed at one of his textbooks. He told me, "In describing the wind firmness of different species of trees, it said, 'White pine is not subject to windfall because it has a good root system. It's not damaged by wind at all.' Well, we found out it wasn't as windfirm as we'd like. It went down quick. It was tall, in fairly dense stands, and once it started to go, it was like a row of dominoes. That book was published in March 1938, and the hurricane hit in September. So much for that idea."

TWO

Tropical Cyclone

Memory is a capacity both individual and cultural. Think back to when the recent economic downturn began in 2007 and how frequently it was compared to the Great Depression. Some called it the "great recession" to reinforce the comparison. Because so many individuals could tell firsthand tales about the deprivations of the 1930s, the memory of it was still very much alive in the public consciousness.

When the 1938 hurricane slammed into Long Island and New England, not a single soul alive had ever experienced a major hurricane in New England. People were dumbfounded. They didn't know what hit them because the last time a storm of anywhere near this magnitude had pounded New England was more than 120 years before, in 1815. The last identified hurricane of any size to hit New England was a compact but intense storm that clipped Rhode Island and eastern Massachusetts in September 1869. Any remnants of hurricanes to rain on New England in the six decades since then were not so different from any strong gale and were not identified as hurricanes. So many generations had lived and died since the last hurricane that any recollection of it had disappeared from the public consciousness, and most people thought Thirty-Eight to be New England's first hurricane. The *Boston Globe*'s lead story the next day began, "A tropical hurricane of incredible violence swept Massachusetts and the rest of New England last night for the first time in the history of the area."

Hurricanes were so unfamiliar that it was possible to err in the other direction, by lumping them in with tornadoes, as the *Burlington Free Press* did in its September 22 edition:

> Some older Vermonters compared the storm with the hurricane of August 6, 1900. In 1900 the storm was brief but much more intense, spending its violence in a half-hour of cyclonic fury that ripped hundreds of trees from their anchorage, tore down wires and poles and inflicted tremendous damage to house tops and buildings. . . . Yesterday's scourge of wind and rain came hurtling in from the south. The 1900 cyclone roared down on the city from the north and northwest. . . . The 1900 hurricane was concentrated entirely on a small section of Lake Champlain and northern Vermont, wreaking its vengeance in a narrow path.

Clearly, the 1900 event was not a hurricane, but in 1938, journalists and the public they served knew as much about atomic bombs as they did about hurricanes—nothing.

Today, New Englanders on the coast have experienced enough high winds and storm surges during hurricane season that it's now an accepted part of the weather calendar. Even inland in central Massachusetts, New Hampshire, and Vermont, the remnants of tropical storms have periodically brought enough rain to flood river valleys and knock bridges off their abutments. We all know hurricanes and probably have a story or two to tell about our experiences with them. So it's hard to imagine a time when people could be shocked at the arrival of a hurricane. But that's how it was in 1938. As far as people in the Northeast were concerned, hurricanes hit the South. If you lived in North Carolina or Florida or Texas, you might be prey to them, but not if you were a Yankee. It just didn't happen.

■

What exactly is a hurricane? For the Taino people of the West Indies, a hurricane was an angry goddess. Their island paradise generously provided a world of abundance and bright sunny days, but each year, the evil goddess Huracan might come and change all that, showing up when summer's heat began to wane. Huracan rode the winds and joined forces with thunder and floods to pummel the islands and lay waste to paradise. Early Spanish explorers, some of whose voyages were plagued by the work of these gods, brought the Taino word home to a world where such mayhem was unknown. Punishing wind, buckets of rain, earth-rattling thunder, and impossible-to-anticipate flashes of lightning breaking through the swirling rain and clouds—indeed, an angry goddess.

Another two-word definition of hurricane less fanciful and much less evocative than angry goddess is tropical cyclone. Each word is crucial.

To qualify as a hurricane, a storm must be born in the tropics and develop its power out over the warm waters of the Atlantic Ocean not far north of the equator. Because of the ocean's huge mass, it doesn't change temperature readily, so it takes until August and September for ocean temperatures to heat up to the 80 degrees Fahrenheit necessary to fuel the hurricanes that reach us. By then the ocean has warmed to 80 degrees not just at the surface but down to a depth of at least 150 feet, and that huge mass of warm water provides the reservoir of energy to help random thunderstorms coalesce into a system. If the disturbance develops a rotation, it becomes known as a tropical depression, generating wind speeds up to 38 miles per hour. Many storms never graduate past the depression stage, but when they do blow more strongly—39 to 73 miles per hour—they are termed *tropical storms*. In tribute to that increased power, the National Hurricane Center provides them names, a protocol that began in 1953. On average, a dozen named tropical storms develop in the Atlantic each hurricane season. Hurricanes differ from tropical depressions and tropical storms only in magnitude. When Tropical Storm Steve develops winds that exceed 74 miles per hour, it becomes Hurricane Steve.

Sadly for me, that won't happen because the naming authorities don't include Steve in their current list of six S names that rotate every year. My only chance to make the list is to have one of the S names retired, which happens when a hurricane is so devastating it must be memorialized. Hurricane Katrina, for instance, will always refer to the storm that devastated New Orleans and the Gulf Coast in 2005. No more Katrinas. The names of only seven hurricanes that battered the Northeast have been retired: Agnes in 1972, Bob in 1991, Carol in 1954, Diane in 1955, Gloria in 1985, Irene in 2011, and Sandy in 2012. Sandy was replaced with Sara, so I missed a golden opportunity.

If your name is Xerxes or Quentin, you're totally out of luck, because there are no named hurricanes with initial letters of Q, U, X, Y, or Z. Instead of twenty-six alphabetical names, there are twenty-one, but only once has the number of tropical storms exceeded that allotment. If we see a repeat of the very active 2005, Greek letters will kick in again, so Alpha, Beta, and Gamma would succeed William or Wanda.

Beyond being tropical in origin, the storm must be a cyclone. Its winds must be cyclonic, blowing in a circular pattern around a center of low pressure. In the Northern Hemisphere, these cyclonic winds rotate in a counterclockwise direction; below the equator, they rotate clockwise. In the Pacific and Indian Oceans, tropical cy-

TABLE I
CATEGORIZING STORMS

Designation	Sustained Winds (miles per hour)
Tropical depression	38 or less
Tropical storm	39–73
Hurricanes	
Category 1	74–95
Category 2	96–110
Category 3	111–29
Category 4	130–56
Category 5	157 or higher

Note: At different stages in a storm's life, its designation will change. If it develops a cyclonic rotation, it is called a depression. Once it reaches tropical storm status, it is given a name that stays with it through its lifetime. The Saffir-Simpson Scale rates hurricanes according to their sustained wind speeds. Category 3 and higher are considered major hurricanes.

clones are known as typhoons or simply as cyclones, and have the same properties as a hurricane.

Not all cyclones are tropical. In fact, the winds around any low pressure storm system are cyclonic. A January blizzard moving eastward from the Great Lakes or a warm, muggy storm bringing rain northward up the Atlantic coast are cyclonic systems, but not tropical in origin. Further, cyclonic storms with winds greater than 74 miles per hour spring up every year in the United States without being hurricanes. They're called tornadoes, and their structure differs completely from that of a hurricane. As fierce as they are, tornadoes are localized rather than widespread, a weather event rather than a weather system. (Just to further confound the picture, not all tornadoes are cyclonic—5 percent of them are anticyclonic, that is, they rotate clockwise.) A hurricane system can spread out six hundred miles, with the worst damage concentrated in the hundred miles or so surrounding the eye. A tornado, on the other hand, will rarely span more than a mile across and more likely has a core more on the order of a football field or two in diameter. Its characteristic funnel cloud reaches from storm clouds aloft to its narrow core, which contacts the earth. Spawned by inland thunderstorms in the summer heat, the whirlwind can travel a short distance or as far as one hundred miles. These intense storms are mercifully brief.

For decades, I have relied on meteorologist Mark Breen and his colleagues at the Fairbanks Museum and Planetarium in Saint Johnsbury, Vermont, for my daily weather forecasts. Thousands of people in Vermont and adjacent New Hampshire, Massachusetts, New York, and Quebec can instantly recognize Breen's resonant voice from his *Eye on the Sky* radio broadcasts. I visited the man with the radio voice in his office outside the recording booth downstairs at the planetarium. I wanted him to help me understand the origins of our everyday weather and of the extraordinary event of 1938.

Breen makes meteorology presentations at schools and for civic organizations, so he has a knack for keeping it simple. He made sure I understood common weather terms, beginning with an explanation of the difference between high and low barometric pressure. High pressure does not mean that the pressure is way up high; it means there's a lot of pressure, a lot of weight, a lot of density. Cold air is heavier than warm air, and what's known as standard atmosphere (a happy medium of cool, calm air at sea level) weighs 14.7 pounds per square inch. Measured in a barometer, a tube within which mercury can rise and fall according to the weight of the air —otherwise known as atmospheric pressure—this translates to a barometric pressure of 29.92 inches. The normal range of readings in a mercury barometer moves only an inch or so in either direction from standard. Unless it reaches an extreme low, forecasters won't mention the specific barometric pressure but will simply report whether it's rising or falling. A falling barometer means that the air coming into the region is lighter than the present air, more energetic, more volatile. The incoming air mass will probably bring clouds, and the clouds might drop some precipitation. Ordinary low pressure accompanying a thunderstorm might be around 28.85. Hurricanes bring much lower readings. The lowest reading during the 1938 hurricane was 27.94 at Bellport on Long Island.

The narrow range from low to high in a mercury barometer can mislead us into thinking that the difference between high pressure and low pressure isn't that great. And that perception wouldn't change by adopting the other common scale, which uses millibars or hectopascals. The hurricane low of 946 millibars compares with the standard of 1013.25. This hurricane low is only 6.6 percent lower than standard, but that represents a tremendous difference in atmospheric pressure. Think of it as the difference in density between a volleyball and a bowling ball.

Air has weight, and air moves from places where the air is heavy to where it weighs less; as the heavy air moves in, the lighter air lifts. I experience this in the comfort of my own home nearly every morning in winter. We heat our house with a woodstove on the main floor and close off the upstairs bedrooms at night, and it can get pretty chilly up there. Sitting downstairs having my morning coffee, I can tell when someone opens a bedroom door upstairs because a burst of cold air comes rushing down the stairway. Cold air is heavy, and as it sinks, the warmer air has the opportunity to rise.

Breen said, "Across the globe, you have higher and lower pressure in various arrangements. There's a net air flow from higher to lower pressure. This net air flow means that around high pressure, air will be moving away from it. With low pressure, air is moving in toward it." Air around high pressure exits and spirals outward in a clockwise fashion. That interacts with the adjacent low pressure systems where the air is trying to come in, creating a compensating area of counterclockwise circulation around low pressure.

Continuing with the big picture, Breen gave me a quick lesson on the general air circulation model of the northern hemisphere by drawing it and dividing it into three zones, each with 30 degrees of latitude. Air is cold in the polar zone and hot near the equator. Cold air sinks, hot air rises, so if all else were equal, air would travel from the polar regions to the equator at the surface and then flow back toward the pole in the upper atmosphere.

But some things are not equal, and one of the most important inequalities is the speed of the Earth's rotation at different places on the Earth. At the equator, it's moving at a rate of 1,037 miles per hour; at the north pole, on the other hand, the speed of its rotation is a mathematical riddle, because at the true point of the axis it doesn't move at all, it just takes all day to make one spin around the pole. Keeping that in mind and planting ourselves at the arctic circle (latitude of 66 degrees), the rotational speed is 457 miles per hour. It's that much slower than at the equator because the circumference at the arctic circle is less than half that at the equator.

By now, Breen was sketching furiously with his pencil. He said, "Because of the spin of the earth, the air at the surface tends to turn instead of going straight. The spinning globe means this air that's trying to head southward from the pole turns to the right, meaning east to west."

That's because of the Coriolis force, which accounts for the difference in speeds on our spinning sphere by dictating that anything moving above the Earth's surface in the northern hemisphere moves to the right of its intended direction of travel. Though it's often called the Coriolis force, it's more accurately the Coriolis effect. There's no true force exerted, and the effect happens because the ground beneath is traveling at a different speed at the destination than at the starting point. So air moving from the polar region to the south curves to the right, which makes it feel to someone on the ground that the wind is coming from the east. Winds and weather systems in the polar region come from the east. Let me pause for a moment to clarify a point already known to many, but perhaps not all: in the parlance of weather forecasters, an east wind means it's coming *from* the east, not flowing toward the east. Same with all the other directions. So any time I refer to wind with a direction, I'm naming the direction it's coming from.

The bottom third of Breen's sketch showed the Hadley Cell, the zone from the equator north to 30 degrees latitude. In the portion of it that rides along the equator, approximately ten degrees north and south, there is no prevailing wind. Nor is there any Coriolis force. These are the doldrums, avoided if possible by all sailors not wanting to spend days adrift with no wind to propel them. Day by day at the equator, warm moist air rises, condenses, and drops rain on the tropical rainforests that occur on any land along the equator: the Amazon basin, equatorial Africa, the Philippines, New Guinea. Buckets of rain accompanying thunderstorms keep the land constantly humid. This constant low pressure zone experiences rain more days than not.

Rising air from this zone heads poleward, but it gets only a third of the way there. It begins to sink by the time it has reached 30 degrees. Breen said, "If you look around the globe at around 30 degrees, this is where you'll find most of the major deserts in the northern hemisphere. The air tends to warm up and dry out, and it's a zone of quiet weather where the air primarily sinks."

At the surface, this heavier air is pulled back toward the equatorial low pressure zone, and in doing so it turns to the right, going from east to west. These are the trade winds, which transatlantic voyagers have harnessed for centuries by sailing south from Europe until they reached these easterlies, which fill their sails and speed them west across the sea.

Polar winds and tropical winds both blow from the east. Balancing these two easterlies are the westerlies in the midlatitude temperate zone between them. The 30-degree latitude line passes through

Jacksonville, Florida, and Houston, Texas, and the 60-degree line serves as the northern border of the provinces of British Columbia, Alberta, Saskatchewan, and Manitoba. New England sits centered in this temperate zone, with Fort Kent, Maine, at 47 degrees and New Haven, Connecticut, at 41 degrees.

Over most of the North American landmass and the Atlantic Ocean to its east, these westerly winds are the counterpart to the easterly trade winds. "That's the general west to east motion that we observe with storms in our region," Breen said. "You have this general motion where you have rising air, which tends to create clouds and precipitation. This midlatitude storm track gives us our weather—it doesn't stay in exactly the same spot but shifts north and south."

If you watch the weather forecast, you'll see the inevitable eastward progression of weather systems across North America. The weather follows the track of the jet stream, a powerful river of air five or so miles up in the upper atmosphere that forms above the undulating line at the Earth's surface where polar air and warmer air meet. The jet stream is pushed north or south depending on the relative power of the competing air masses, and its position directly influences the surface winds beneath it.

So we have an established pattern of easterly winds in the polar and tropical regions sandwiching westerly winds across the temperate center. The tropical easterlies and the temperate westerlies tell us a lot about everyday weather patterns in the zones affected by these winds, but they do more than that. These winds determine the path that most hurricanes take because they help to steer it. It's hard to imagine that a force as powerful as a hurricane can be at the mercy of the relatively gentle prevailing winds, but that's the way it works. Breen said, "You might think that air mixes together quite readily, but oddly enough it doesn't." It's as if air masses were solid rather than gas, so the hurricane follows the direction of air flow.

■

Lourdes Aviles, who teaches meteorology at Plymouth State College in central New Hampshire, comes from the land that Huracan ruled, having been born in Puerto Rico in 1970. In 2013 she published *Taken by Storm, 1938*, which examines the meteorology of Thirty-Eight and the challenges it presented to the U.S. Weather Bureau's forecasters. I visited her in her office on campus, an office whose furnishings must put students at ease. On your way in, you walk beside a bookcase that might make you wonder whether you've entered filmmaker Tim Burton's office by mistake. Its top shelf is populated by an impressive collection of figures taken from one of Burton's stop-motion animation films, *The Nightmare before Christmas*. When you're greeted by the Pumpkin King, Oogie Boogie, and Dr. Finklestein, you know you're not in the office of a stuffy academic. On another wall hangs a sculpture of a cloud festooned with three brass spigots. Each spigot bears a label designating a certain amount of rain: Drizzle, Showers, and Cats and Dogs.

Aviles told me, "I grew up in the Caribbean, so every year there were hurricanes that were trying to come our way. Sometimes they got there, sometimes they didn't. Once in a while a storm came close enough to be felt, but when I was a child none of them caused a lot of devastation." When she pronounces devastation, she does so with Latin syncopation not routinely heard in northern New England.

Though Aviles grew up in a period of relatively inactive hurricane seasons, as each season approached she heard tales from her grandparents of powerful hurricanes, especially the crippling Category 5 hurricane known locally as San Felipe that left half a million people homeless on the island in September 1928. Memories of those stories and those storms came to her later when she was in graduate school at the University of Puerto Rico, working on her degree in physics. "This was in 1995, and it was a very active hurricane season after all those quiet years," she said. Her assignment was to put together a seminar on any topic of her choice, and as the rains fell and the wind howled, she decided to do her seminar on hurricanes. "I enjoyed it so much—doing the research, presenting it—and it was like a light bulb over my head: 'I don't have to keep going with physics.' It was getting more and more theoretical, and it was not really why I was interested in the first place. I just wanted to understand how things worked. It had taken me years to figure out that this was the direction I wanted to go." The next year, following her newfound passion, she came to the U.S. for a Ph.D. in atmospheric science at University of Illinois and has been teaching at Plymouth since 2004. Sparked by stories of San Felipe and bemused by local mythology that correlated a bountiful avocado crop with an active hurricane season, she has become an expert on African easterly waves and their role in forming tropical storms from what might otherwise be just an innocuous cluster of thunderstorms.

Any hurricane whose impact is felt in the Northeast has been born over the Atlantic Ocean in the tropical latitudes, with many of them developing off the coast of Africa. Hurricanes are born over the warm ocean and die out when they cease to be fed by warm ocean air. What they do during their lifetimes—whether they kill people, destroy bridges, and flatten forests or spend their days simply churning up the ocean that feeds them—depends partly on chance but mostly on the weather systems around them.

For a tropical depression to organize, it needs a combination of high humidity, a deep mass of warm ocean temperatures, and light winds. Hurricane season technically runs from June through November, but the requisite warm, humid conditions are most often in place toward the end of summer into fall, so August, September, and

October are prime months. Aviles explained to me how a thunderstorm or a cluster of them becomes organized, which means that it develops a cyclonic circulation. The Coriolis force initiates the counterclockwise rotation as air approaching the organizing low pressure from any direction is deflected to the right. The net effect of all of this air rushing into the system and turning to the right is a counterclockwise flow that spirals into an ever-taller chimney sucking up the moist air being fed into it. This rising air feeds the clouds and further perpetuates itself by reducing the atmospheric pressure at the surface.

A hurricane sucks up moisture through the chimney of the eye and by doing so dramatically reduces the air pressure at the Earth's surface. Think of your blender making a smoothie. Above the blades, the center is hollow, relatively calm, and all the churning is going on along the walls, the perimeter. While your blender is powered by electricity, the energy of a hurricane is self-perpetuating once it gets started and as long as the conditions for that energy are met: constant introduction of warm, moist air at the hurricane's base. "As air spirals toward the center, it ventilates the surface of the ocean and collects the heat and vapor that evaporates from the surface," Aviles writes in *Taken by Storm, 1938*. This churning of the ocean brings deeper water to the surface, and if that water were any cooler, it would reduce the overall temperature and stifle the energy. In fact, the passage of a hurricane churning up the water beneath it reduces the prospects for a subsequent hurricane to pass in its wake because the mixing cools the water in what is known as a cold wake. "Heat increases the air's buoyancy, causing it to further accelerate toward the center, collecting even more heat, further increasing its buoyancy and so on," Aviles writes. "One can think of a mature hurricane as an engine converting the heat energy provided by the ocean to the mechanical energy of the hurricane winds." The chimney becomes the eye of the storm, which in a mature hurricane can extend sixty

33

thousand feet—more than eleven miles—into the atmosphere and can span fifty miles.

Born in an area with little prevailing wind, the storm can feed itself for days without moving much. At some point, it will come into contact with the easterly trade winds, which will steer it across the ocean toward the Caribbean and North America. Even though a hurricane at this stage could feature winds exceeding 100 miles per hour, the lateral progress of the system tends to be slow because the trade winds steering it typically don't exceed 15 miles per hour.

The trade winds skirt the Bermuda High, known to Europeans as the Azores High. The span between those two sets of islands is two thousand miles, which gives us an idea of the magnitude of this semipermanent high pressure system. Every summer, this elongated dome of high pressure sits out over the Atlantic, and the storms follow the easterly trade winds south of it. The clockwise flow exiting the high pressure system keeps the storm sailing westward. West of the high, the clockwise air circulates to the northwest, then north, and many storms follow this path. As they do, hurricanes will typically come in contact with the generally west-to-east air flow that prevails over the continent, and be steered to the northeast without making landfall.

Meteorologists call this recurvature, as the hurricane's track boomerangs back northward and then northeastward. The steering effect of the Bermuda High and the westerly continental air flow explains why a relatively small percentage of hurricanes make landfall in North America. To do so, conditions have to be just right. Aviles cites the track of Hurricane Bill in 2009 as the prototype for the typical recurvature and the path that forecasters in 1938 expected for the Great Hurricane.

Data from the National Oceanic and Atmospheric Administration (NOAA) for the period from 1968 to 2011 show that an average of a dozen tropical depressions develop high enough wind speeds to

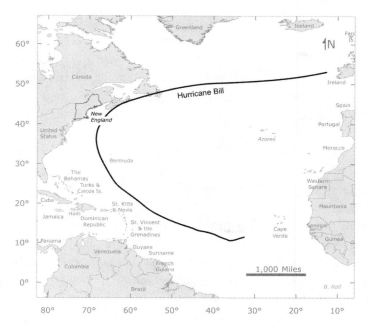

The track of Hurricane Bill in 2009 serves as the prototype for the classic recurvature pattern. Bill formed west of Africa, tracked west toward the Caribbean islands, then curved northward, and continued curving to the northeast into the North Atlantic. Brian Hall, Harvard Forest, Petersham, Massachusetts

be designated as tropical storms, with half of them reaching hurricane status. Fewer than two (an annual average of only 1.7) hurricanes make landfall in the Gulf Coast or Atlantic states. Beginning in late August, people all over the eastern half of the United States stay alert to the tropical storms as they form and gather steam, even though chances for a destructive landfall are so slim.

Using time-lapse satellite imagery, NOAA creates videos that show the storm tracks over the course of a hurricane season. (You can view "The 2011 Hurricane Season in 4.5 minutes," for instance, at the NOAA Visualization Lab's website.) The sequence of these tropical depressions beginning and ending illustrates elegantly the interplay of the hurricane systems with the continental weather systems. What we see is a dynamic situation with endless permutations. Most of the time, the track takes some sort of a turn to the right and heads into the North Atlantic, where the colder waters don't provide enough fuel for the hurricane's furnace, causing many storms to peter out. Sometimes the storm does not recurve and continues to head west through the Caribbean and into the Gulf of Mexico. That was the path taken by Katrina in 2005, as well as the deadliest hurricane ever to hit the United States, the 1900 hurricane that hit Galveston, Texas, killing seven thousand people.

Since European settlement, interior New England has been hit by only three devastating hurricanes: in 1635, 1815, and 1938. Other hurricanes have brought rain and wind to the area, but if our criterion requires a tropical storm to affect central Massachusetts with hurricane force winds, we're limited to these three. Loosen the parameters to include coastal New England and it's a different story. High winds and torrential rain from tropical storms more frequently have hit southern New England, Cape Cod, and the islands of Martha's Vineyard, Nantucket, and Naushon and the other Elizabeth Islands. This region conveniently pokes its chin out into the Atlantic, and when you lead with your chin, you get walloped once in a while. It bears repeating, though, that for many years these gales were not recognized by the populace as hurricane-related events.

Note that New York (with the exception of Long Island, which geographically relates more closely to New England than it does to the rest of the state) has never experienced a Category 1 hurricane. Even Sandy, which devastated parts of New York City and was given the new hyperbolic title of Superstorm, had been reduced to post-tropical cyclone status by the time it made landfall in New Jersey, eighty miles south of the city, on October 29, 2012. That said, its strong winds, heavy rains, and storm surge had been felt onshore for many hours while Sandy was offshore and still a hurricane.

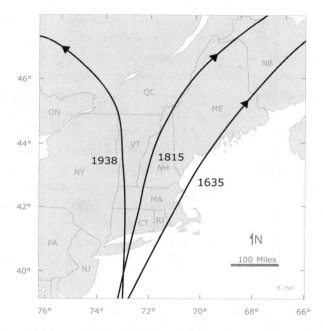

The three hurricanes that have caused severe damage in interior New England occurred in 1635, 1815, and 1938. Each sped forward at more than 40 miles per hour. As these depictions of their tracks show, they maintained their strength because they did not make landfall south of New England. Brian Hall, Harvard Forest, Petersham, Massachusetts

Next in line behind Cape Cod and the islands, Connecticut and Rhode Island have the most experience with hurricanes. If for some reason you wanted to live in a place that holds the amenities of city life along with the best potential to be knocked down by a Category 3 hurricane, you could do no better than choosing Providence, Rhode Island. Sitting at the head of Narragansett Bay, Providence's downtown was submerged by a tidal surge of sixteen feet in 1938, a result of the hurricane's arrival coinciding with high tide. If you were a city planner, you could look at that storm surge as a fluke, or you could build sea walls for protection. Providence chose the latter, but only after Hurricane Carol in 1954 slammed them again with a storm surge nearly as powerful, disproving the fluke theory. Providence gets hit regularly, but not like in 1938.

As we move farther inland, Massachusetts is less likely to be hammered. Measured from south to north, Connecticut is only seventy miles wide, but that distance has generally been enough to dissipate a storm's energy. Springfield or Northampton, Massachusetts, can expect to be hit by a Category 1 storm every 53 to 120 years, according to the National Hurricane Center. Add the sixty-mile span of Massachusetts to the track a hurricane must take, and Vermont and New Hampshire are even less likely to be nailed. The twin states can expect a Category 1 storm every 121 to 290 years.

To reach Vermont and New Hampshire with full force, a storm has to avoid spending a lot of its energy over land, so it has to take an almost perfectly diabolical path to make it up the Connecticut River Valley. That's just what it did in 1938.

A lifetime ago, when I was playing Little League baseball, one of our most biting taunts was "He can't hit the broad side of a barn," which described the pitiful inaccuracy of an opposing pitcher's throwing arm. Well, from the looks of it, North America qualifies as the broad side of a barn—from out in the Atlantic, all you've got to do is aim west and you can hardly miss it.

But most hurricanes do miss the inviting target of North America, and almost none of them hit the much smaller strike zone of the Northeast with full force. Those that make it to New England generally do so in a much-diminished state because they arrive only after having spent considerable time over land, which dissipates their energy. As devastating as 2011's Hurricane Irene was for much of Vermont, the damage came entirely from the rain, not from the wind, which had been played out by having made earlier and episodic landfalls at Cape Hatteras, then southern New Jersey, and finally New York City. If New England's soil hadn't already been saturated from a remarkably wet summer, Irene's rains could have been absorbed and they would have drenched but not destroyed. Paul Sisson, of the National Weather Service, made a presentation to meteorologists that both Breen and Aviles told me about. Sisson compared the Vermont effects of Hurricane Floyd in 1999 with those of Irene. The two storms followed nearly identical tracks and dumped similar amounts of rain, but Floyd didn't raise havoc because, unlike Irene, it came after a dry summer. The woods were perfectly capable of absorbing the five inches without turning streams and rivers into torrents.

Baseball provides a particularly apt analogy for storm tracks. A right-handed pitcher throws a curveball by snapping his wrist to impart a clockwise spin to the ball. The righty's curveball naturally moves from right to left. If a hurricane were a righty, and the pitcher's mound were a spot in the ocean a hundred miles east of Cape Hatteras, it would be easy to drop a curve ball across Long Island, into the mouth of the Connecticut River, and right up the valley. But because of the prevailing continental westerlies, a hurricane is in effect a lefty, and a left-handed pitcher's curveball, with its counterclockwise spin, moves from left to right. If the hurricane aims too far to the left, it hits land well south of New England, which dissipates its energy. If it successfully avoids Cape Hatteras, the Jersey Shore, and any other tempting target and stays out to sea, by the time it reaches New England's latitude, it has been pushed out into the North Atlantic.

The arc of most hurricanes that reach the Northeast follows the path of a lefty's curveball. We get soaked by rain—sometimes days of it—as the storm passes to the east. When its path diverges from that pattern, it's because other conditions have fallen into place perfectly to keep it from being blown to the east. That's what happened on September 21, 1938. The storm passed Cape Hatteras and then disappeared from forecasters' view. They expected it to recurve to the east, but it kept heading north. It maintained this unwavering northward thrust for the length of the eastern seaboard, from Jacksonville, Florida, to Bellport, New York. It set its eye on Long Island and didn't waver, traveling at an unprecedented 50 miles per hour. Everyone was shocked, including the National Weather Service, when it reappeared on Long Island. It lost hardly any speed after it made landfall, and its path took it north and then northwest. Instead of a curveball, Thirty-Eight was a tailing fastball.

A look of wonder brightened Mark Breen's eyes when he said, "What's unique about the hurricane of '38 is it's the only major hurricane in four hundred years of European settlement here to have come through New England across Vermont and into eastern New York state. That was a very unusual path."

Here's how it happened. The Bermuda High spent the summer of 1938 extended farther north than normal and was stronger than normal. This unusual position may have contributed to New England's abnormally rainy summer by causing storms connected to the jet stream to cover the same ground repeatedly. By September each year, the summer heat is dissipating, and some early incursions of polar air head southward, pushing the jet stream to the south. This extension of a mass of low pressure to the south is known as a trough, and in the days leading up to September 21, the trough was unusually strong for that time of year, and it reached all the way from Canada to the Carolinas and Georgia. A trough this deep and powerful is common in late fall and winter but unusual in September. Aviles writes, "Consequently, hurricanes and such deep troughs do not commonly interact."

Until it began to recurve, the hurricane had been surrounded by warm tropical air, but now it came in contact with two opposing air masses of unusual strength. Timing is everything. Breen told me, "If there wasn't a hurricane entering the picture, this situation with the two strong systems would have worked itself out over three or four

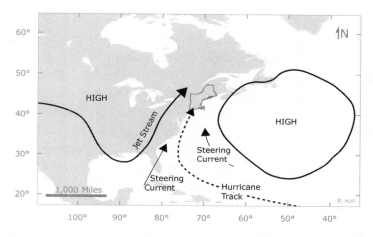

The meteorological setup for Thirty-Eight steered it on its unusual track. The Bermuda High was stationed uncharacteristically far north in the Atlantic, preventing the hurricane from recurving back into the Atlantic. Clockwise rotation of winds around the high combined with counterclockwise rotation from the deep trough in the jet stream to create steering currents that drove the hurricane straight into New England. Brian Hall, Harvard Forest, Petersham, Massachusetts

days. It would have flattened out. The hurricane came at the time when the two of them were at their extremes."

A ridge of high pressure over the ocean faced off against a trough of low pressure over the continent. The counterclockwise flow caused by the trough to the hurricane's west and the clockwise circulation exiting the high combined to steer the hurricane right up a corridor running almost due north. With steering currents this strong, the hurricane could not possibly move out to sea.

It made landfall as a Category 3 hurricane, and then gradually made the transition to an extratropical cyclone. An extratropical cyclone is simply a cyclonic low pressure system; a nor'easter, for instance, is an extratropical cyclone. Its energy comes from the presence of strong temperature gradients that stir up the winds and generate clouds and precipitation, so it is characterized by weather fronts. Meteorologists now know that it's fairly typical for a hurricane to transform into extratropical when its warm-water energy source dissipates. Nearly half of all hurricanes make this transition, including Sandy and Irene. Aviles said, "When they become extratropical, they expand and they become asymmetrical, too. And they weaken because the energy source is not as strong." Instead of being a tightly organized circle surrounding a core, the eye collapses on the trailing side, and the storm clouds take on a shape that Aviles likens to a comma. The transition is not a sudden stark difference but rather takes place gradually. And as Thirty-Eight made its transition, it didn't weaken dramatically because it had a replacement source of energy in the tremendous contrast in air temperatures in the opposing fronts.

"The timing of the transition could not have been worse," Aviles wrote. "The storm made landfall as a major hurricane, with its intense winds and strong storm surge near the center. At the same time it was starting to transition, thus expanding its wind field as it moved inland." Focused and concentrated as it plowed into Long Island and southern New England, it spread out with winds that were slightly weaker but still fully capable of knocking down whole stands of trees. This one-two punch maximized the damage.

Besides the rarity of Thirty-Eight's track, another anomaly was the storm's rapid forward speed, which also resulted from the unusual setup. Aviles has shown that its legendary forward speed— pegged at 60 miles per hour or more in many accounts—has been exaggerated. Still, once it started heading north, it averaged 50 miles per hour even after landfall, an uncommonly fast pace. Most hurricanes tend to poke along at less than 20 miles per hour, and it can take days for the huge system to work its way through any particular spot on its path, especially if it transitions to extratropical. Hurricane Irene was a slow mover like that, averaging 15 miles per hour for most of its life, giving it plenty of time to dump six to eight inches of rain over a twenty-four-hour period onto an already saturated Vermont. In contrast, Thirty-Eight sprinted from a position east of Cape Hatteras to the south shore of Long Island in only seven hours. Its center crossed Long Island at 3:30 P.M.; Hartford, Connecticut, at 4:30; the Vermont border at 6:00. High wind preceded the center's arrival and continued after it passed, but the storm blew through relatively quickly, thus the aptness of one of its early monikers, the Long Island Express. The wind in any one locale lasted no more than four and a half hours and peaked for an hour or so. All of the wind was gone from Vermont by 11:00 that evening, and in Vermont newspaper accounts the next day, it was referred to as "last night's storm."

At any particular place in the hurricane's track, it came and went in less than five hours, but those five hours have stayed in the memory of anyone who experienced it. Sandy and Irene were damaging storms, but Thirty-Eight matched their storm surges, rain, and floods and blew them away with 100 mile per hour winds across the landscape.

THREE

Disparate Destruction

New England had been inundated by tremendous floods in the recent past—the deluges of November 1927 and March 1936 persist as the benchmarks for measuring high water in New England—but these had come from extended periods of rain (along with snow-melt in 1936) that overwhelmed the rivers' capacity to move water. They were an overabundance of the normal. The hurricane of 1938 seemed to have nothing normal about it.

On September 22, people within a fifteen million–acre footprint woke up to a scene of catastrophe. Stricken was a ninety-mile-wide swath from the Long Island shore to the Quebec border that included almost all of Long Island, Rhode Island, and New Hampshire, and substantial parts of Connecticut, Massachusetts, and Vermont, and even a chunk of western Maine.

The hurricane hit the coast just before a high tide bolstered by the approaching new moon. The high tide was just the ground floor of a gale-driven storm surge that brought salt water to places that had never tasted it. Fifteen feet of water swamped streets, its salt bath ruining thousands of cars and trucks. It picked up moored yachts and fishing boats and beached them on their sides a quarter mile inland. It splintered houses and cottages and swept them off their foundations. Entire beachfront communities disappeared, sucked by the undertow back into the ocean. The surge undermined bridges and roads and picked up railroad tracks as if they were Lionel train sets in the hands of a tantrum-fueled six-year-old. Telephone and power lines snaked through yards and streets. The damage to infrastructure along the coast of Long Island, Rhode Island, and Connecticut was catastrophic.

Damage from flooding was not confined to the coast, as dams on upstream rivers and ponds were breached, and bridges were knocked off their pilings in Massachusetts, Vermont, and New Hampshire. The 1927 and 1936 floods had inundated many of the same areas and had prompted the construction of flood-control dams. Some of these did their job in 1938, keeping a bad situation from getting worse. The Connecticut, the Merrimack, the Thames, and many of the rivers flowing into them neared or exceeded previous highwater marks. The Ware River in central Massachusetts flowed six feet higher than it had in 1936, ripping apart houses, factories, and the fire station in the town of Ware. The devastation to the region's infrastructure would end up requiring repairs costing $300 million in Depression-era dollars, approximately $5 billion today.

The transformation of the Category 3 hurricane into an extratropical storm widened the scope of the damage. The relatively tight footprint of the hurricane expanded as does a bullet that contacts resistance. Few people, whether urban or rural, escaped loss altogether. In New England, flooding was a well-understood peril, but the wind was an entirely new nightmare.

Nearly all of the wind damage happened to the east of the storm track. That disparity came courtesy of the wind's two originating sources, the storm's counterclockwise rotation and its forward motion. Let's look at these two forces separately at first. Picture a hurricane standing stationary with its eye centered over the single point on the bank of the Connecticut River shared by New Hampshire, Massachusetts, and Vermont. Downstream Northampton would experience wind from the west while upstream Bellows Falls, Vermont, would feel wind from the east. Bennington, Vermont, would have winds from the north, while Nashua, New Hampshire, would feel them from the south. All of these winds would be of the same 80 mile per hour velocity, but a circle of calm with a radius of twenty-five miles would surround that point.

Hurricanes aren't stationary, of course, and as we know, the 1938 hurricane was traveling nearly due north at a 50 mile per hour clip. Because a hurricane is a cyclonic system, its counterclockwise winds can either enhance or diminish the effect of its forward motion, depending on your position relative to the path. The most vulnerable position in any hurricane is to the right of the track (east in this case). On that side, the strongest winds—combining the 50 mile per hour forward motion with the 80 mile per hour rotational winds— blew steadily at as much as 130 miles per hour and gusted at half again that rate. Thirty-Eight's highest wind speeds were recorded on an anemometer at Blue Hill Observatory on a hill in the Boston suburb of Milton, Massachusetts, nearly 100 miles to the east of the storm track. There, sustained winds from the southeast reached 121 miles per hour and gusted at 189 while ripping apart the observatory's best measuring device.

On the other hand, locales to the west of the storm track saw the forward winds more or less canceled out by the rotational winds blowing from the north. Land on the right side of the storm track experiences more wind, the left side has more rain. This right-left differential occurs with every hurricane, but if a storm is poking along at 15 miles per hour, the difference is not pronounced. Thirty-Eight's rapid forward speed exacerbated the disparity.

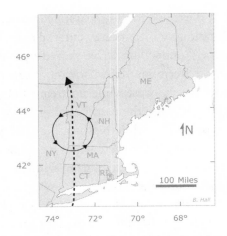

The strength of winds experienced on the ground during a hurricane depends on the combination of two forces. The forward movement of the storm, which was 50 miles per hour in 1938 at landfall, combines with the cyclonic rotation of winds around the center. On the right side of the storm track, the rotation adds to the forward motion, creating wind in excess of 100 miles per hour. Winds on the left side of the track are considerably less, because the cyclonic wind negates that of the forward motion. Brian Hall, Harvard Forest, Petersham, Massachusetts

Any location would experience a shift of wind direction as the storm approached, reached its same latitude, and then blasted on by. The directional shift would be most stark in those cases where the eye went directly overhead. The eye is calm, clear of any of the storm clouds that spiral through the rest of the storm, and it experiences the lowest barometric pressure. Surrounding it and containing it is the eye wall, the ring of the most violent rain and wind.

Reports from people who experienced the eye tell us that as the storm approached them, the wind blew from the southeast. The leading eyewall brought its brutal winds, but when that passed by, they experienced a period of calm. Meteorologist Charles Pierce reported that on Long Island the calm center was experienced as far west as Brentwood and as far east as Mattituck, towns that are forty-three miles apart, confirming that the eye had at least a forty-three-mile diameter. In the calm center the wind was so slight that a cigarette could have been lighted without difficulty. The sun even broke through, and people thought that the storm was over. When the wind returned an hour later, it was from the southwest, but in the trailing wall it was not as strong as the leading wall.

The eye was estimated to be fifty miles wide at landfall, and those locales that saw only part of the eye pass over would have a shorter interval of calm and a sequence of slightly different wind directions. As the cyclone transitioned to extratropical, the center lost its symmetry, which may explain the puzzling fact that people in New Haven reported the eeriness of the calm center but residents of Hartford—situated only thirty miles to the northeast, they should have spent some time within the eye—didn't report the calm.

To the west of the storm track, rain embedded in thunderstorms continued to soak places that had already experienced plenty of rain from the prehurricane low pressure system. But there the wind wasn't howling so madly, so any shift in direction might have gone unnoticed. Locations east of the track got pummeled with the one-two punch of rain and wind. As the storm sped along on its northward track, anyone to the east of it experienced a shift in the wind direction that started from the east, maybe even the northeast. Then, as the storm approached, the wind would grow stronger and shift to the southeast and then south. At a certain point in the storm's passage, the gusts became strong enough to start knocking trees over. In most locales, this happened during the period the wind was coming out of the southeast.

Recall fourteen-year-old Fred Hunt's experience under the white pine tree straddling the road in Rindge, New Hampshire. All of the trees thundered down around him in a quick ten-minute burst. After that the wind kept wailing, but all the damage was done. The punishing winds in Rindge came from the southeast, as they did over

Foresters cruised the countryside tallying the amount of blown-down timber.
These two, dressed in fedoras and ties, were working in the White Mountain
National Forest in Gorham, New Hampshire. United States Forest Service

much of the fifteen million–acre swath of destruction to the east of
the storm track.

Twenty miles southwest of Rindge in Petersham, Massachusetts, is a research forest owned since 1907 by Harvard University.
Harvard Forest encompassed 2,100 acres in 1938 and has grown
to 3,750 acres today. Until the hurricane hit, the staff was actively
managing its timber stands to produce income to fund the research
at the forest. They saw it as a home to research but also as a model
forest, a place for demonstrating to the public how diligent forestry

could benefit everyone, from the owners of the woods to the lumber industry and the various workers who depend on that industry. In truth, the forest hadn't been self-sufficient despite the directors' best efforts. Still, with backing from alumni supporters, it continued its forestry research and efforts to grow valuable timber.

For decades, the forest's founding director, Richard Fisher, and other foresters had tended and thinned their stands of old-field pine while planting newly abandoned pastures and hayfields with white pine, red pine, and white spruce to showcase silviculture to other private owners. Blasted by the gale, the tall pines rocked back and forth in the gusts, and they couldn't withstand it for long. Some pines snapped off below the crown, but most of them were wrenched from their perch and lay in a tangle, their huge rootwads vertical, their trunks horizontal—not the way anyone wants to see carefully tended trees.

File cabinets in the Harvard Forest archives are filled with inventory and harvest data taken in the thirty years before that September afternoon when all their efforts to grow superb pine were rendered to splinters. When a research forest loses 70 percent of its mature trees, what choice do its directors have but to try to turn the lemons into lemonade and shift their attention to a study of catastrophic disturbance to forests? In residence that autumn was a new graduate student named Willett Rowlands, who was quickly enlisted to document the extent to which the various stands had been blown down. With his clipboard and camera in hand, he scrambled through blowdowns and got himself covered with oozing pine pitch day after day as he documented the damage on each of the plots, noting the species, direction of fall, and the degree of damage. He visited nine hundred mapped stands, and his painstaking work would become the subject of his thesis for his master's degree. Subsequent grad students soon followed in his tracks and described the nature of the forest that was regenerating in the woods that had been blown down and salvaged.

These investigations provided a wealth of data that has been well used by many researchers since then.

Successors to Fisher, including Hugh Raup, who studied the cultural history of the landscape, turned their attention back to using the forest primarily as a site for research. And when David Foster, a forest ecologist with a passionate interest in the interplay between cultural history and natural disturbance, became director in 1990, he took the university's forest in the direction of understanding the long-term processes—both natural and cultural—that shape New England's forests. Today it hosts botanists, soil scientists, and ecologists from across the globe as a field site for long-term ecological research.

Long, lean, and still youthful after twenty-five years at the Harvard Forest helm, Foster manages—despite his administrative and academic responsibilities—to spend lots of time outdoors, always with his camera and its heavy zoom lens slung over his shoulder. Given his interest in large-scale environmental change—he previously studied fire in the boreal forests of Labrador—Foster found much of interest in the rich history of land use and disturbance in the forest he now oversaw. Over the years, he and his colleagues at Harvard Forest have published a number of scientific papers that go a long way to providing a clear picture of the ways that hurricanes have damaged New England forests over the last four centuries. And even though forest ecology is the prime research focus, the Harvard Forest researchers had to develop a sophisticated understanding of hurricane meteorology and the physics of wind.

Let's start with the trees. The extent of the blowdown depended largely on what kind of trees—which species and how tall—were in the crosshairs. Starting from Rowlands's data, Foster documented that conifers were more vulnerable than hardwoods, and that white pine—the region's tallest and fastest-growing tree—was the most vulnerable of all. As a rule, the older, taller trees were more susceptible than shorter, younger trees, but in the case of pine, even individual trees only fifteen years old were knocked over. Entire stands of thirty-year-old pines were flattened. Hardwoods, which in general have a stronger root system, weren't blown down unless they were taller, which meant they would have to have been considerably older than the fast-growing pine.

Whether conifer or hardwood, the older the tree or stand of trees, the more likely it was to uproot rather than snap. Picture a tree as a lever that has one end anchored in the ground. The taller the tree the longer the lever and the greater force it can exert on the ground where it's anchored. Older forest-grown trees tend to be tall, so they transfer more of the wind's force to the ground even as they bend. As trees increase in diameter, they lose their suppleness. It then becomes a question of whether the trunk is stronger than the roots, and in 1938 the roots snapped much more frequently than the trunks. Less than 15 percent of the pines broke rather than being uprooted, and an even smaller percentage of hardwoods snapped off.

In forest-grown trees, competition reduces the vertical extent of the trees' crowns. That's because in a natural, tightly spaced stand, most trees prune themselves of the lower branches that sunlight doesn't reach, so limbs, branches, and leaves cover only the top quarter or third of the tree. Along with competition comes a sort of cooperation, as forest-grown trees rely heavily on their compatriots to retain their upright position. The swaying trees work almost as a single organism, each tree absorbing a small portion of the wind's force. But when faced with a gale that exceeded anything before, and wind coming from the direction opposite from normal, the stand lost its capacity to stay upright. The dominoes started falling.

In many cases, the tallest tree went first. The huge white pine that saved Fred Hunt's life blew down before the others because it was taller and more vulnerable. The wind speed at canopy height—60 to 120 feet in mature forests, depending on the species mix—exceeds that at ground level, where friction slows it. A forest canopy is not a smooth surface because some trees excel at reaching the sunlight and dominate their cohorts. In mixed stands, some species grow taller than others, with white pine the usual winner. White ash will normally poke its head above sugar maple or American beech of the same age. These disparate crown heights roughen up the canopy's texture and provide a hook for the gale to grab the tallest trees.

Foster revisited another observation from Rowlands's thesis and explained why so many of the old-pasture trees survived the wind even though they were standing in the open, taking the full brunt of the storm. In an open-grown tree, the crown starts low and extends for a much longer percentage of the tree height, so the wind's leverage is greatly reduced. In engineering terms, the turning moment is reduced. Having spent a lifetime fending for themselves in the face of periodic wind, they developed broader and stronger root systems. In an interesting variation on this, many of the original seed trees for stands of old-field white pine withstood the tempest within the forest they'd spawned. They had established themselves in the open, developed huge multitrunked crowns, and produced copious amounts of seed. Windfirmness is not heritable, so their forest-grown progeny lay spread out on the ground around them.

Emery Boose was David Foster's coauthor on some of the papers. In the 1980s, Boose was handling Harvard Forest's computer systems, and the two began talking about ways to analyze historical data by mapping them through the newly available geographic information system (GIS) that merged mapmaking and aerial photography with powerful computer tools, such as databases and statistical analysis.

They used GIS to analyze the patterns of damage in one of the heavily damaged areas of Harvard Forest, known as Tom Swamp. Besides species and tree height, exposure to the wind was the other most important factor determining what blew down and what didn't.

The sprawly form and the prominent branch stubs show that this white ash was a pasture shade tree. It survived Thirty-Eight despite its vulnerable exposure to the southeast wind. All the smaller trees have grown up since the farmer stopped grazing cows here. Stephen Long

In interior New England, hurricane winds are strongest from the southeast. That's true for Thirty-Eight and for any hurricane past or future because there are only four different paths a hurricane can take and have an impact on New England. Two tracks involve earlier landfall to the south, either up through the Gulf States from the Gulf of Mexico or landing between North Carolina and New Jersey and heading north. The third track doesn't make landfall at all, passing near Cape Cod out in the Atlantic. In each of these three scenarios, the entirety of New England can get drenched. The only way for it to be flattened is by the fourth track, crossing Long Island and heading due north. That's the only way for any hurricane to deliver powerful winds to central Massachusetts or beyond.

Thirty-Eight's center passed thirty-nine miles to the west of Petersham, so Harvard Forest was close to the eastern eye wall. That was a position primed for damage, but Foster pointed out that not all Petersham was equally smashed. Protected areas—those in the lee of any of the hills—received less wind and suffered little damage.

Boose created a GIS-based model of protected and exposed sites. In the absence of any strong topographical features—hills or valleys —all areas to the east of the storm track would be similarly exposed. Gently rolling terrain doesn't change the equation much. But when the elevational contrast is built up, some sites become more exposed to the wind and others more protected. The steeper the slope, the more it's exposed.

Given that the strongest winds were coming from the southeast, any hill that opened itself to the south or the east wore a bull's-eye. Geographers designate a hill's aspect using compass bearing. A slope that faces due east has an aspect of 90 degrees, while a south-facing slope has a 180 degree aspect. In interior New England, slopes with aspects between 90 and 180 degrees are in the most vulnerable position to hurricane winds. The backsides of these hills—assuming they fall away to the north or northwest—might escape altogether, particularly if pitched steeply away. Back slopes just below the crest would experience some of the wind, but the 1938 storm didn't venture abruptly down a lee hill.

As Boose became more interested in analyzing wind patterns and damage, and he had more computer power at his fingertips, he began working on a computer model that could reproduce the effects of historical hurricanes. In a 1994 paper, he and Foster introduced two computer models, HURRECON and EXPOS. With enough historical documentation of damage on a town-by-town basis, he could create a simple model of the storm track and speed, the timing of arrival in any location, and the wind speed and direction for that location over the duration of the storm. Boose said, "There are certainly random events in a big storm like the hurricane of '38 but there are patterns on a regional and landscape scale that seem to be reasonably consistent from one storm to another that can be modeled."

The 1994 paper focused on a recent hurricane—1989's Hurricane Hugo, which pummeled Puerto Rico—and the 1938 hurricane. Testing the model on contemporary storms like Hugo, he could verify the accuracy of the parameters because complex meteorological data are now quickly available from the National Oceanic and Atmospheric Administration (NOAA) about any weather event.

"Models are always approximations so we always have to bear in mind their limitations," Boose said. "HURRECON uses the location of the hurricane itself, and that estimate has uncertainty to it even in modern times. There are uncertainties about exact location of storm center, for instance. And the model has assumptions about uniform wind distribution, which of course aren't quite right. HURRECON might not differentiate accurately the damage in one town versus an adjacent town. That kind of scale is pushing the accuracy, but over the larger scale, the county scale, it ought to do pretty well."

The EXPOS model takes it one step further. By bringing in GIS topographical overlays, it predicts relative exposure and protection for local sites. Boose speaks softly and deliberately in sentences far clearer than those most people can write, and he explained why they developed the models. "Our especial interest was not in the meteorology per se but in the impact on the forest," he told me. "Forests grow very slowly, and forest dynamics unfold typically over a period of decades or centuries. Trees and forests live out much longer lives than humans do, so not only is their future longer than our own, but their past exceeds our memory. Many of the trees that we see every day were established long before we came to know them, so the long-term records kept at Harvard Forest are exceptionally valuable in helping us to understand how the present conditions came about. To really understand today's forests we need to find out as much as we can well into the past."

Boose's two models help explain the wind patterns in a particular place, allowing him to reconstruct with some certainty what happened. Earlier, I showed how the winds shifted when the storm swept through. This scenario of shifting winds plays out anywhere within the footprint affected by the storm, with more dramatic shifts happening closer to the eye wall. For Petersham, HURRECON predicted—if that's the correct term for something that happened seventy-five years ago—that the winds would begin more easterly and that the peak winds would come from 135 degrees. Foster and Boose have confidence in the accuracy of their wind modeling, so they needed to interpret their data that showed that the average fallen tree had been laid low by wind at 124 degrees, 11 degrees east of what they expected.

Foster said that stands with shorter trees (less than forty-five feet tall) did fall in the predicted direction, but there were more trees that were taller, and the majority of these taller trees toppled from east winds before the wind reached its peak velocity. The shorter trees were slightly more windfirm and held on longer because they provided less leverage. Overall, the trees fell in a relatively narrow range of orientation, which suggests that most of them went down as soon as the wind reached catastrophic wind speed. Gusts generally come in the same direction as the sustained wind, and gusts are what knock trees over, though trees' purchase may be made precarious by the sustained winds at lower velocity.

Harvard also owns a twenty-acre inholding in Pisgah State Park in Winchester, New Hampshire. Harvard's Pisgah forest is an old-growth stand of huge white pines, but almost all of the pines have been horizontal rather than vertical since 1938. Unlike the extensive salvage that followed the storm in Petersham, not a single stick of wood was salvaged at Pisgah. It stands as a laboratory and a monument to what happens in a forest that people don't directly influence. Pisgah was closer to the storm track. In fact, it was just within the radius of the eye, so HURRECON predicted two peaks of wind, one each from the leading and tailing eye wall. Treefall orientation showed that all of the pine at Pisgah went down in the first peak wind of the leading eye wall.

The Harvard Forest research paints a thorough picture of the ways its trees were damaged. And we can apply their findings and their model to forests farther north in Vermont and New Hampshire, though those states' hilly and even mountainous terrain made wind directions and speeds more variable locally. As the storm reached farther inland, its winds diminished some, though not dramatically, because its forward momentum kept it alive.

So we know what species and age of trees were most likely to fall. And we know that if those trees grew in a protected pocket, even though they might have been tall enough to blow down, they would still have been standing the next morning. And we know that there were inexplicable local deviations to the broader patterns they have identified.

This is the most intriguing conclusion, that the damage was extremely variable. Anyone taking a long look at the local damage could convince himself that the entire forest was destroyed and that the only trees left standing were the ones that were snapped off and not uprooted. On the other hand, if you were to fly over New England in a plane low enough to provide a good overview of the destruction on the ground, you'd see a mosaic of damage, with many shades of gray between the extremes of flattened and untouched.

Some large patches were flattened, others had more than half their trees blown down, but in some forests, the trees were only inconvenienced by having their leaves stripped. And some stands escaped damage altogether. Trees were blown down in 904 townships across fifty-one counties in six states. More than thirty thousand families woke up the next day to damaged forests. That's a big loss for a tremendous number of people. The 600,000 flattened acres were in patches ranging in size from one-tenth of an acre to 90 acres, with a distinct tendency toward the smaller sizes. Assuming that Harvard Forest's damage distribution was representative of the region as a whole, 80 percent of the openings were smaller than 5 acres.

■

For seventeen years, I edited and published a magazine called *Northern Woodlands*. Beginning with its first issue, forester and naturalist Virginia Barlow wrote a column that each time focused on an insect or disease that plagued a particular tree. These villains tend to have fanciful names—white pine blister rust, gypsy moth, and shoestring root rot, for instance—and the list has recently included nonnative killers, such as Asian long-horned beetle and emerald ash borer. After seventy columns, Barlow still hadn't repeated herself; the list of organisms plaguing trees could very well prove to be endless. In the woods, life walks hand in hand with death: death from wind storms, from insects and diseases, from ice storms, from logging, even occasionally from old age. Unless you look for it, you can live unaware of the processes going on in the canopy above you, on the ground beneath your feet, and even the area within easy reach as you walk through the forest. Once you begin to notice death, it's all around you. The most beautiful woods you've ever walked through are full of dead and dying trees. It's the nature of life in the woods. Death begets life.

Death is normal, and necessary, and natural. That goes for humans, for woodland animals, and for plants, including trees, which live longer than animals and take longer to disappear. As the largest and longest-living organisms in the forest, trees have the largest influence on the lives of all the others. Caterpillars feed on tree leaves, birds feed on caterpillars. Ants feed on dead wood, pileated woodpeckers feed on ants. Mice and moles thrive in the protected environment of rotting trees on the forest floor. Without an ever-replenishing horde of these small prey animals, the predators—foxes, fishers, hawks, and owls—would have a hard time making a living. The leaf litter on the forest floor supports microbes and larger decomposers at the bottom of another food web. And so it goes. All of this complexity begins with the food and shelter provided by trees.

On average, natural mortality might take 1 percent of the trees each year. The death of mature trees allows for the natural regeneration of new trees because their death creates a gap in what might otherwise be a full canopy. The gap allows sunlight to reach the forest floor. The agent of death and the way the tree dies combine to determine how large a canopy gap is created. An individual tree that wastes away from disease or insect damage will leave a gap equal to the size of its disappearing crown. Its neighbors will eventually spread into the newly available canopy space, and seedlings of species that tolerate shade might also seize the brief opportunity to establish themselves.

Snow can damage tree crowns if it comes when the trees aren't dormant, either in the early fall or the late spring. Huge limbs and branches snap from the weight of wet snow collecting on early or late leaves of deciduous trees. Injuries from ice are similar, but ice can wreak havoc even in dormant trees, weighing down branches and limbs so they break off. Weather conditions have to be just right for ice to accumulate and linger on a tree, but when it does, it can skeletonize the crowns of many trees over a wide area. Serious ice storms in 1998 and 2008 smothered trees across large stretches of the Northeast. Usually, trees don't succumb immediately from broken crowns, but the resultant stress renders them vulnerable to an infestation.

With windstorms, the range in damage is wide. If it's mild in speed or duration, it might simply break off susceptible branches and limbs. If it blows harder or is sustained for longer, it can uproot a tree or a whole swath of trees. When a single dominant tree is blown down, it can create an opening twice the size of its height because it will take other trees down with it. At a height of eighty feet, a tree's fall could initiate a gap of a quarter acre, roughly the size of a typical urban house lot. Microbursts formed within summer thunderstorms can easily knock down multiple patches of an acre or so.

Where does fire fit into this catalogue of natural disturbance? Even though every summer brings news of catastrophic fires out west, northeastern forests are not highly prone to fire damage. The northern hardwood forest, which covers substantial parts of New York, Vermont, New Hampshire, and Maine, has been dubbed "the asbestos forest" because its inherent moisture makes it so difficult to burn. Oak-dominated central hardwoods of southern New England are more susceptible, especially when they have a component of pine. Coniferous forests are the most fire prone, though our focus today on fire suppression keeps the acreage burned artificially low. It's another natural but rare disturbance.

Every forest in the Northeast experiences frequent small-scale disturbances that create relatively small gaps in the forest. Interspersed with these are infrequent disturbances creating larger gaps. This plays out in a steady loss from insects or disease taking out a tree here and there augmented every so often by wind, snow, or ice with a broader destructive reach. Decades can pass in this way, and when the decades pile up to become centuries, chances increase that a hurricane or fire will enter the picture and take down an entire stand of trees. At the start of this chapter, I wrote that the floods

Strong wind blew down this clump of mature hardwoods, creating a long but not particularly wide opening in the forest. Some seeds will probably germinate in the opening, but the crowns of the adjacent trees will fill in the gap within a decade or two. The excavated pits paired with the adjacent large rootwads will form the pit-and-mound topography characteristic of blowdown. The remnant snow in the lower right-hand corner lies in a pit formed by a previous blowdown. Stephen Long

of 1927 and 1936 were an overabundance of the normal, and that the hurricane of 1938 seemed to have nothing normal about it. Yes, that's the way it seemed to someone who looked with anguish on the ruins of his house, barn, or woods. In truth, however, hurricanes of this magnitude are normal. They are just exceedingly rare.

Meteorologists expect major hurricanes to return to Rhode Island and Cape Cod every 33 to 52 years, to the rest of southern New England every 53 to 120 years, and to northern New England every

121 to 290 years. Many generations will probably live and die in Saint Johnsbury, Vermont, or Plymouth, New Hampshire, without a major hurricane. On the other hand, it could come next summer. That's the nature of it.

David Foster and his colleagues at Harvard Forest have shown us that not only did Thirty-Eight blow trees down at the broadest catastrophic level, it also created smaller gaps. Seventy-five percent of the mature trees at Harvard Forest blew down, which is not the same as saying that it took down all the trees in three-quarters of the Forest's twenty-one hundred acres. Instead, one-third of the acreage experienced the most severe damage. The largest of these nearly complete blowdowns reached ninety acres, and there were ten more jackstraw holes in the woods larger than twenty acres, but the preponderance of damage was in patches smaller than five acres, with some gaps as small as one-tenth of an acre.

In five hours across a footprint of fifteen million acres, Thirty-Eight made Swiss cheese of the canopy. Knocking down thousands of entire stands and perforating the canopy in hundreds of thousands of small patches, it created more gaps in those five hours than might normally occur over the course of fifty years of ordinary disturbance.

The most recent preceding major hurricane had hit in 1815 with similar power and a similar storm track. Its path, however, was through a New England that was mostly cleared of trees by the settlers who had been expanding inland from the coast since the middle 1600s. By 1815, 60 percent of Massachusetts and Connecticut was growing grass, corn, and wheat instead of oak, chestnut, and hickory. And much of the forest had been cut over, in most cases more than once. Consequently, there wasn't the same destruction in the forests.

Human industry had created a situation that diminished the impact of what we believe to have been a Category 3 hurricane. And human inaction—letting much of that farmland revert to a forest filled primarily with white pine—left it in a very precarious position in 1938.

■

The woods I have owned in central Vermont since 1988 had their own bout with Thirty-Eight, which I will detail in Chapter 9. Our woods are situated at the end of what's shown on USGS topographical maps as Hurricane Ridge, and indeed, the wind can howl up there.

One summer afternoon in 2002, a freak windstorm known as a microburst blew down thirty thousand board feet of timber in these woods. I'd been at the magazine office, and it had started like any summer thunderstorm with distant rumbles, the occasional flash, off-and-on rain. Suddenly the wind blew the daylight away. A barrage of rain pounded the metal roof. The four of us in the office jumped up to close all the windows against it, but we missed the open top half of a double-hung window. Horizontal rain found the opening and sprayed an interior wall sixteen feet away and everything between. Doors slammed shut, loose paper flew everywhere. The roof vibrated wildly, struggling to stay attached. I heard a pop from the transformer on the power pole, and all of our computer screens and LEDs went dark. Two trees in the backyard crashed down, but a huge landmark pine somehow stood its ground. And then the wind eased back on the throttle. It had done all of its damage within a couple of minutes, and the four of us gathered to express awe. None of us had ever experienced such a sudden blast of wind.

In the magazine business, nothing can be accomplished without a phone and computer, so we all left work, wishing each other luck in reaching home. My normal five-minute commute home took nearly an hour because I kept coming upon trees across the road. Eventually, I found a way through the maze, though its last obstacle—the pines blocking our driveway—could not be avoided. I cut them with a chainsaw, and some friends pulled the branches, limbs, and short logs off to the roadside. Progress was slow, one cleared tree at a time. Some were bare trunks, the trees tall enough that their crowns cleared the road. That made it easy. More difficult were the pine tops sprawled in the road, because each limb in a tangle is under either of two powerful forces, tension or compression. Released from the force holding it, a branch could spring back and swat you like a fly. Or it could release a log to roll and pin your leg to the gravel. Implode or explode, which would it be? Avoiding either takes patience, and I cut slowly as I read each limb's response to the changing physics. The saw's roar precluded conversation but begat impromptu sign language that grew more comical as we approached the end. Branch by branch, tree by tree, we cut away the dozens of trees the wind had knocked down across the quarter mile of road to our house.

Without the chainsaw, it would have taken all day. Instead, we cut our way through it in an hour. Beyond the blowdown, the road was open for a ways. Then, just short of the crest of the last hill, a large spruce crossed the driveway and blocked the view of the house. I hopped out of the truck and skirted the spruce on foot. There would be lots of time to cut it later, and I had to see whether any trees had hit the house.

One pine sprawled twenty feet from the house, the nearest miss. I recognized that as a blessing, but there were so many trashed trees, including some special ones. Red oaks might be a dime a dozen farther south, but our woods have exactly three of them. Two of the three had been hit by falling trees. The first, a ten-foot-tall sapling, was pinned to the ground by a fallen pine. The second, taller and stouter, leaned at a 45 degree angle propping up a large uprooted aspen. When I freed the oaks from their burdens, I bent the sapling back upright and staked it. The other oak—less supple at five inches in diameter—barely rebounded, but the danger of it uprooting from the aspen's weight was gone.

I explored the woods above the house, and I began to feel that Mother Nature had a particularly cruel streak. First, the oaks, and now my only elm. Very few American elms survive in the woods past a six-inch diameter. You'll notice them roadside the year they die, their leaves wilting in early summer when the beetle carrying Dutch elm disease invades them. The only one extant in our woods—tall and straight with a twelve-inch diameter—had escaped the beetle for decades but now had been wrenched to the ground.

Hundreds of trees—half of them white pines, the other half hardwoods, particularly sugar maple and white ash—lay on the ground. What a great way to learn about root systems. The roots of well-established trees don't form a ball despite the nursery term *rootball*, nor do they show any sign of having a taproot like a carrot. Instead, they are shaped like a thick dinner plate. The tangle of roots had hoisted up topsoil, heavy boulders, and slabs of bedrock embedded in them. The scraggly perimeter of these upright rootwads showed snapped-off roots.

I hiked through intermittent mayhem. The damage was in patches, none larger than half an acre. A flicker, my favorite woodpecker, flew up a woods road ahead of me with its characteristic dips and rises, undulating until out of sight. That's the way the wind had blown, dipping down here and there but skipping large stretches in between. Where the gale touched down, it laid the trees down all in the same direction, the weight of one causing the next to tip over, until the last falling tree landed between trees rather than on one.

This microburst, with its blast of wind as high as 100 miles per hour, is the closest I've come to a hurricane. Of course, it's ridiculous to compare a miniature to the real thing, but the summer microburst gave me an inkling of what it means to be overwhelmed by an intense and unforeseen act of God. Our wind blew for two or three minutes; Thirty-Eight raged for five hours, most fiercely for an hour. Each of the fifty or so of my neighbors hit by the sudden storm experienced serious loss; in 1938, thirty thousand people walked out into their woods stunned by what they saw. We cut our way through the tangle with a chainsaw, which wasn't introduced until the late 1940s. That was a decade too late for the thousands of people wielding axes and crosscut saws to open up the roads and reconnect to the outside world.

FOUR

Better than Box

On September 21, 1938, the power of nature overwhelmed thirty thousand owners of woodlands, thousands of stranded motorists and train travelers, and hundreds of thousands of people who lived or worked near the coast. These people came face to face with a force so much more immense than any they'd ever seen or imagined, a force of such magnitude that its pounding surge registered on the seismograph at Fordham University in the Bronx as if it were a major earthquake.

The roaring wind changed the way people look at the natural world. I've spoken to dozens of octogenarians who remember the day as if it were yesterday. Whether they were four or fourteen, most of them speak first of the incessant scream of the wind. Any time Thirty-Eight comes up in conversation, they live through the wind all over again. Jim Colby, of Boscawen, New Hampshire, was a young man of eighteen at the time. "The wind began to pick up gradually. When we got home the wind was blowing hard, but nobody could comprehend it was going to blow harder," Colby told me.

"And then all at once the skies opened up and the wind came and all the trees in our dooryard went down." He paused for a moment. "Unless you were there and saw it, you just can't comprehend it."

The experience of all of these people brings me back to my junior high studies, where we were told about one of the primary themes in literature, "Man versus Nature." It was positioned among other conflicts individuals found themselves engaged in, for instance "Man versus Man" and "Man versus Self," the latter of prime interest to any cringing adolescent. Man's bout with nature propelled short stories like Jack London's "To Build a Fire" and Stephen Crane's "The Open Boat." Later, many of us read novels built on that theme: *Robinson Crusoe*, *The Old Man and the Sea*, more Jack London, and the grandest of all, *Moby-Dick*.

In each of these stories, a character comes up against nature in its impassive, imposing power. In the Yukon or in a storm at sea, the protagonist faces a dire situation in which no help can be assumed, and he must struggle to see whether his strength, his wit, and his skill can surmount this supreme challenge. In these modern tales, nature is not personified in the gods as the Greeks did, or as the Taino people did with Huracan. Neptune didn't cause Crane's shipwreck, nor did he roil the seas that toss the open boat about. Still, as an adversary, nature takes on characteristics that seem almost like a personality, and readers caught in the spell of these tales may wonder whether nature is actively hostile or merely ambivalent. That same question must have been voiced by the people whose lives were ruined by the raging storm.

Under those circumstances, it could be easy to forget that humanity's relationship to nature is not simply adversarial. The flip side of disaster is reliance. Everything we use every day comes from the natural world. Not only those items that feel natural, like our food, clothing, and shelter, but also our decidedly synthetic goods— vehicles and electronic devices—start with ingredients that are animal, vegetable, or mineral. Whether we participate directly in the procuring of these materials through hunting, gathering, fishing, farming, logging, or mining, or whether we purchase our goods through the invisible network that brings them to the door, we are sustained entirely by materials from Earth's core, surface, and atmosphere.

Nature both destroys and sustains us. But what about our impact on nature? With technological advances that increase our work output and speed, humans now wield so much collective power that we influence and alter the natural world in profound ways. The conservation movement that began in the late nineteenth century and contemporary environmentalism decry what we have done to the Earth in the service of progress. Not all of our changes are as stark, immediate, and negative as dropping bombs in wartime or removing mountaintops for coal. Water shortages, loss of species, and degraded habitat spring from our overuse of nature's fruits. Similarly, our passion for growth engenders a level of consumption that is fueling the changing climate. E. O. Wilson, the originator of the concept of biophilia, recently said about the ongoing threat of species extinction, "We are a biological species living in a biological world. We need the rest of the living world."

Gordon Whitney suggests in his brilliant history of postsettlement change in the New World, *From Coastal Wilderness to Fruited Plain,* that our understanding of humanity's role needs to be recalibrated. Contemporary environmental histories concentrate on the ecological degradation that transformation has caused, while nineteenth-century historians extolled our advance into the New World as transforming a wasteland into a garden. "Where does the reality lie?" Whitney writes. "Most human activities have beneficial as well as deleterious effects. Deciding exactly what is beneficial or deleterious for a given ecosystem alone is a question which requires a great deal of scientific thought." And since we are undoubtedly part of that ecosystem, progress we've made in fighting disease, improving standards of living, and increasing human life spans needs to be part of the equation. He pairs a quotation from John Muir—"The early settlers regarded God's trees as only a larger kind of pernicious weed"—with one from a lesser-known agricultural college dean: "We have these farms, these citizens, these railroads, and this civilization to show for it, and they are worth what they cost."

Our dependence on nature is absolute: we cannot survive without tapping into its bounty. Nature, on the other hand, could do just fine without us, though for the moment at least, she seems stuck with us. In 1922, William B. Greeley, chief of the U.S. Forest Service, wrote, "The progress of civilization has been called a struggle between human wants and natural resources. And no part of this age-long contest has been more clear cut than the effort of mankind to supply its need for wood." With that in mind, let's take a look at the series of human-induced changes to the environment that set the stage for the forests of Massachusetts and New Hampshire to be pounded so mercilessly by Thirty-Eight.

■

From our vantage point in the early decades of the twenty-first century, purchasing rural land continues to be a good investment. For at least half a century and until 2007, when land values began to drop across the country, most people have assumed that the only direction land value could go was up. For years, real estate agents have been telling prospective buyers of land, "They're not making any more of it, you know." And to give that credence, sales have indeed rebounded from the temporary fall, and town clerks are again busily recording deeds. The development potential of rural land once again is sparking the speculative impulse.

But for most of human history, rural land has been valued entirely for its capacity to grow a crop, not for its development value.

Until the early 1960s, a logger would evaluate a woodlot's timber and offer to buy the land (not just the timber) at a price based on what he could recoup in an immediate harvest. Similarly, farmers looking to add acreage made their calculations based on the parcel's crop-growing potential. In the decades following 1850, when Massachusetts farmers stopped milking cows and shearing sheep and traveled to town to make a living, it took faith and imagination for them to hold onto that fallow land and continue paying taxes on it.

I described in Chapter 1 the process by which the land was initially cleared and then abandoned. The wholesale change in the landscape accomplished by pioneering generations of land-clearing farmers was gradually reversed in the mid-nineteenth century as the fields filled up with trees. Animals responded quickly to the changing habitat, and as with any change, there were winners and losers. Bobolinks, meadowlarks, and other grassland birds lost their preferred habitat, so their populations declined. Other species that had been absent or rare expanded into the burgeoning shrubland, attracted by the thick low cover of the shrubs, seedlings, and saplings. It has been said that John James Audubon never heard a chestnut-sided warbler sing. No doubt it's apocryphal, since *Birds of America* contains a plate with a male and a female of the species. But if he had lived beyond 1851, he would have become familiar with these shrubland specialists, which rebounded along with the indigo bunting, chipping sparrow, and ruffed grouse.

Once it's established, white pine grows quickly, so in each reverting field the shrubland habitat was ephemeral. The old-field pines grew so quickly that they became a tall, full-canopied forest within forty years. The process was initiated at different starting points all across the landscape. It transformed a centuries-old checkerboard of farm fields interspersed with woodlots that had been cut periodically but never cleared for farming. By 1890, someone walking "cross lots" from home to town would have passed through a much more diverse arrangement of farm fields, shrubland, young forest, and older primary forest. A checkerboard had become a mosaic.

Even though still young, the pines were large enough to cut and turn into lumber by the 1880s. The colonial clearing of the land had been accomplished by individuals on a local scale. In the 1850s, starting in New York and Pennsylvania and spreading first to the Lake States and then to the South and the Pacific Northwest, logging and sawmilling became an industrial endeavor. Lumber flowed like rivers out of the Lake States' pine belt. To build the cities and towns of the Midwest, nearly 300 billion board feet were harvested in those states by 1900, which dwarfed the volume that had been cut throughout New England, and it had only taken sixty years. Mythical logger Paul Bunyan's exploits originated in logging camps at this time, and the scale of the harvest was truly Bunyanesque. The highest single year's production was 1889, when 9.4 billion board feet of white pine was cut, 7 billion of which came from Michigan, Wisconsin, and Minnesota.

Meanwhile, the second growth stands of central New England pine became so plentiful and the volume of timber so impressive, that it triggered the development of a New England white pine industry. New England's pine was in demand because the virgin pine that had grown so prodigiously in Michigan, Wisconsin, and Minnesota had all been cut. This time around, New Englanders cut trees commercially, not just to get them out of the way for farming. Those who had bought or held onto abandoned pastureland turned out to be shrewd investors. Without anyone tilling, repairing fences, or tending cattle, the former farmland produced a timber crop all on its own. And it was a remarkably valuable crop. The value of the standing timber had grown to as much as five hundred dollars an acre in the forty or fifty years it took the pines to reach cuttable size. David Foster and John O'Keefe, both longtime observers of the intersection of the cultural and ecological, point this out in their book about the famous dioramas at Harvard Forest, *New England Forests through Time.* They describe the white pine boom. Between 1890 and 1920, fifteen billion board feet of second-growth white pine was felled, sawn into lumber, and used for boxes, pails, and all sorts of other items that capitalized on its light weight and workability. A steady flow of old-field pine fed this vibrant industry. The pine boom, an accident of forest succession and fortuitous timing, made it profitable for people to own rural forestland.

The turn-of-the-century marketplace needed white pine, and as fortunate as this situation was for the landowners of Massachusetts and New Hampshire, it could have been better if the trees had better form. The pioneer seed trees standing one by one in old fields grow in a sprawly, multitrunked manner, which earned them the name cabbage pine. The trees they spawned didn't win any beauty contests either. The perfect lumber tree grows gun-barrel straight, and its lower limbs drop off from lack of sun as early as possible. That produces a series of straight sixteen-foot logs with inches of clear wood accumulating outside the early branch stubs. Precious few of these treasures grew in the untended old-field stands. "Knotty pine" has somehow taken on a certain cachet in one of those brilliant marketing ploys that turns defects into benefits. It has been adopted by hundreds of motels, restaurants, and other tourist traps—not just for their paneling but as their name—in search of that authentic down-home feel. In truth, the fewer knots the better.

In pine lumber, red knots are acceptable but black knots aren't. That differentiation isn't a color preference but a reflection of the tendency for some knots to loosen and fall out of a board. These black knots come from branch stubs that were long dead when the tree grew over them. Red knots don't cause as much of a problem because they come from live branches that continued to grow with the tree and remained an integral part of the wood.

Knots diminish pine's value, but crooked wood compromises it even more, and old-field pine is particularly crooked because of its attractiveness to the white pine weevil. This native weevil feeds on the topmost shoot of a white pine, starting early, when the pine reaches two or three feet tall. It prefers trees growing in full sunlight, as the old-field pine invariably does. The weevil regards all of the side branches as mere hamburger and instead feasts on the tenderloin of the leader. Adults lay eggs in the central shoot, and when the larvae hatch, the first thing they do is eat their home. This kills the central leader, which triggers the tree's strategy of having one of the branches in last year's whorl abandon its horizontal ways and take over the role of the leader by growing vertically.

The detour that the trunk takes at this spot shows up in the first year, and the swerve never goes away. Worse, two branches sometimes duke it out for the right to be the leader, and the bout ends in a draw, with both forks reaching for the sky. The new leaders may in turn be deposed, and the tree can find itself with twenty or even thirty terminal shoots all contending for dominance. That will generally happen only with the original spotty collection of pioneers in a field, because thicker stands of pure pine limit each tree's side branching. Even in tightly clustered trees, the weevil might cause three or four kinks in each trunk. Over the years, as the forks continue to expand in diameter, they encroach on one another, which weakens the tree's structure, making it likely to split in a high wind. White pine weevil is not kind to its host.

The bad news was that the old-field white pine was plagued with serious structural defects. The good news was that the trees were big and there was a seemingly endless supply of them. Entrepreneurs like Ansel Dickinson saw that the plentiful pine could function perfectly well in making boxes, and he began making pine boxes at a manufactory in Leverett, Massachusetts, as early as 1875.

Dickinson's startup company with its twenty employees evolved into the New England Box Company, with more than fifteen hundred employees in 1930. It owned thousands of acres in Massachu-

This three-trunked white pine was repeatedly affected by the white pine weevil in its first twenty years. Its scraggly form would relegate it to the pulp pile today, but in the days of boxboard, manufacturers could have used most of the tree. Stephen Long

setts and New Hampshire and had fourteen factories within easy reach of its wood supply. This was not merely a homegrown New England enterprise. It was the biggest player in an industry that gobbled up the largest share of white pine in the country. Its corporate headquarters was housed in the Woolworth Building, one of New York City's first skyscrapers, whose neo-Gothic architecture gave it the nickname "the cathedral of commerce."

Commerce relies on packaging, and in the days before plastics and corrugated cardboard took over, pine boxes held fruits, vegetables, fish, raw tobacco and finished cigars, ammunition, shoes, soaps, typewriters, canned goods, any liquids shipped in bottles, and both maple sugar squares and tins of maple syrup. Agricultural New England had transformed itself into industrial New England in the decades surrounding the Civil War, and it shipped its goods in boxes and crates made from local pine.

The boxboard industry was a perfect marriage of strong demand for a product and plentiful supply of the raw material used to make it. Making a box didn't require a long, wide, straight board. Box manufacturers would buy the defect-ridden boards (some of them only six feet long) and resaw them in all three dimensions: length, width, and thickness. They could thus cut out loose knots and crooked grain that was likely to split. Powered by steam and featuring mazes of belts running bandsaws, planers, chopsaws, and nailers, these highly automated plants shipped out boxcar loads of boxes and shooks. A shook was essentially a prefabricated box bundled flat for easier storage prior to assembly by the consumer. New England Box pioneered glued-up corner-lock joints, which precluded nailing and cleating to reinforce the corner. Recognizing that a shipping container's value depended on its strength per weight, the company also relied on the Linderman, an ingenious multifaceted machine made in Muskegon, Wisconsin. It could cut, shape, and join two boards lengthwise with an end-to-end dovetail that needed no glue and was said to be stronger than the raw wood itself.

New England Box was the largest but by no means the only manufacturer to thrive on the old-field pine. Smaller operators competed throughout the region. Meanwhile, a market for circular rather than rectangular containers had also grown up. The Fessenden Company of Townsend, Massachusetts, was the largest manufacturer of kegs and barrels for dry goods, known as slack cooperage to differentiate it from the tight cooperage that could hold liquids. Coopers like Fessenden worked around the problem of knots by buying lumber cut from trees that had rapid height growth, which meant that each year they grew eighteen to twenty-four inches of clear wood between the annual whorls of knots. (Later, Fred Hunt spent much of the 1950s supplying the Fessenden Company with fast-growing pine he cut as thinnings from plantations.)

World War I provided a boost to the already booming industry, but there were troubling signs that the white pine boom was over even as the overall economy rebounded from the postwar recession in the mid-1920s. Richard Fisher, Harvard Forest's founding director, was commissioned to survey the wood-using industries of New Hampshire in 1925 and Massachusetts the following year. At the time, New Hampshire and Massachusetts ranked first and second in production of white pine lumber, but their output had diminished significantly. In New Hampshire, 754 million board feet were sawn in 1907, the peak of the New England pine boom. By 1925, the harvest had settled in at an average of 300 million, only 40 percent of the cut in the boom years.

This was the year that F. Scott Fitzgerald's *The Great Gatsby* was released and the *New Yorker* magazine began publishing, each chronicling an age noted by its wealth and opulence. If ever an industry so inextricably linked to consumption were to flourish, it should be now. But the affluence so visible in Manhattan and the Hamptons was nowhere to be found in Keene, Springfield, and Rindge.

In Boscawen, Jim Colby's father, Joe, charged six dollars a day for himself and his two mules. If he and the mules were rolling roads for the town, he upped it to seven dollars a day. Jim Colby has his father's work ledgers from this time, and they provide a clear look at the economy of rural New Hampshire. Joe Colby was a burly man—he stood six-foot-two and weighed two hundred pounds—who was a good provider for his wife and five children. A farmer with an entrepreneurial spirit, he would often assemble a crew of men to take on large projects. If someone needed his silo filled, Joe Colby brought together ten men and three teams of horses or mules to make it happen. He bought and sold livestock, traveling as far east as Prince Edward Island and as far west as Wisconsin. At a time when brucellosis, known at the time as Bang's disease, was causing contagious abortions in cattle, Colby developed a reputation as a reliable source of Bang's-free cattle.

He also owned a portable sawmill, and he was listed in Fisher's 1925 survey as one of the nearly four hundred timber operators who had sawn at least fifty thousand board feet in 1925. That's a fairly low threshold, since a good sawyer could saw fifty thousand board feet in a week with the portable mills of that day. And while these mills were termed portable, it would take a full week to move one. They weren't the tidy Wood-Mizers you see today. Moving from one woodlot to another would require separate trips for the engine with its eight-foot diameter flywheel, the steam boiler with its twenty-foot smokestack, and the circular saw and its carriage. Workers had to disassemble the timbers and the roofing so they could rebuild it at the new mill site.

A sawyer wouldn't move the mill unless he had plenty of logs to saw. In a typical job in 1921, Colby sawed 443,000 board feet for a man named Dodge, and sold the lumber for $14 per thousand board feet. The job took about eight weeks, and after paying Dodge $2,215 for the logs, and wages to his sawyer, roller, and sticker totaling $2,400, and $541 to move the mill, Colby cleared about $1,000. If the job was big enough, there was money to be made. If not, it was hardly worth the move.

In 1926, Joe Colby finished a sawmill job on Knowlton Hill in Boscawen. He reluctantly moved the mill for a small job up to Northfield, fifteen miles away. Years later, Colby's son Jim told me, "There was a guy who had 100,000 [board feet] of pine that he wanted sawn. Back then you know, there was no money. We always had plenty to

Joe Colby's first mill, set up here on Knowlton Hill in Boscawen, New Hampshire, in 1926. Colby had to retrieve this mill from a woodlot where it had been sitting for a dozen years to begin sawing timber blown down by the hurricane in Rollins Park in Concord, New Hampshire. Courtesy Colby Lumber

eat and to wear, they were always hand-me-downs, of course. But there was no money to do much. My father moved a steam mill, which back then wasn't no easy matter, to saw just 100,000 board feet." After Colby had gone through all the trouble of setting up the mill, it took about two or three weeks to saw the lumber, and when he finished, Colby returned to his other work. He sold a hog for $25, seven bushels of potatoes for $12, and four calves for $31. He hauled wood with his two mules for $6 a day. And for the next twelve years, he never sawed a board. Trees grew up around his saw-mill, which he had left standing on the Northfield property.

In his two reports on the health of the industry, Richard Fisher wrote, "The standing timber is small and knotty, both as compared with the timber in the West and South and as compared with the earlier stands which were cut in New England." By 1925, the biggest and best of the second-growth stands had already been cut.

Trees were sold from the forest in what is still called woods-run. A potential buyer of the timber or the land would make an evaluation of whether the wood's quality was "better than box, box, or poor box," and make his offer based on that. No differentiation beyond that was made in the quality of the individual trees—a lump sum payment bought the lot of it, the bad with the good. Back then, when the forest was cut, it was clear-cut. The buyers "logged it off" with no regard to the forest's future value. Many if not most trees were cut before they had reached their economic maturity. Forest owners were selling their adolescent pine, which was no smarter than a farmer selling a steer before fattening it up.

The lumber was sold log-run. Nobody eyeballed individual boards to sort out those of higher grade for sale at a higher price to better markets. Logs were sawn straight through on portable mills that often produced boards of inconsistent thickness. After slicing off two opposing outer slabs (which were conveniently burned in the steam boiler to power the mill), sawyers cut the log straight through. They didn't turn a log on the carriage in search of the best boards, as is common today. In a practice known as round-edged sawing, the boards' edges were left raw, the bark still attached. Woods-run logs produced log-run lumber. Nothing special, run of the mill.

That was a mistake at a time when lumber sawn from virgin timber and shipped from a great distance—southern "hard pine" and several western species—was competing well with white pine. Not only could shippers find boxes made from alternative woods, they were increasingly drawn to the lightweight corrugated cardboard, which had been around since 1900 or so but was just starting to gain a share of the market. "The fact is that the business has suffered a severe blow, and the consensus of opinion among box men is that it is a permanent one," Fisher wrote. He noted that a number of the smaller manufacturers had gone out of business recently.

At about the same time that Fisher was so cautionary about the prospects for central New England's pine, it was being touted as a resounding success by the U.S. Forest Service and its chief, William B. Greeley. He thought white pine in New England set a fine precedent for what could be expected of second-growth timber. After the Lake States' timber was gone, the lumbermen's attention went to the virgin timber in the South and Northwest in the early 1900s. In the 1922 *Yearbook of Agriculture*, Forest Service chief Greeley (then as now, the Forest Service is a division in the Department of Agriculture) wrote a harrowing account titled "Timber: Mine or Crop?" He made the case that virgin timber had been and continued to be mined, and that it instead needed to be treated like a crop that was nurtured.

From its inception, the U.S. Forest Service has had a decidedly evangelical bent. American forestry was still in its infancy at this time, having been brought to this country from Europe in the 1890s. Gifford Pinchot, conservation adviser to Theodore Roosevelt, founded the Yale School of Forestry in 1900 and became the Forest Service's first chief in 1905. Unlike most branches of study, forestry was founded specifically as a solution to a problem, that human progress requires an enormous amount of wood. Greeley, the third chief, wrote,

Though many of us may never buy so much as a single piece of lumber, every man, woman, and child in the country uses wood every day of his life. Forest products are consumed in obtaining nearly every raw material, and again in virtually every process of manufacture, movement of commerce, and activity of trade.

83

Every ton of steel requires the consumption of wood in mining the iron ore and in mining the coal used to make the steel. All food is produced with the aid of wood in some form, and most of it is shipped in containers made from wood. In short, the general public pays a large part of its bill for wood in the disguised form of the cost of food, clothing, and other articles that contain no wood at all. Ninety-eight per cent of our rural dwellings are of wood. For urban dwellings the percentage is from 59 to 98, varying from state to state. Wooden houses are the easiest, quickest, and ordinarily the cheapest to construct. This has put decent homes within the reach of millions of people, many of whom could not have afforded brick, stone, or concrete structures. Our American living standards are therefore essentially tied in with adequate wood supplies. Wood scarcity carries a universal menace.

In a paper published in 1925, Greeley continued the theme: "The shortage of available virgin timber is growing more and more critical every year. It cannot be emphasized too strongly that the seriousness of this problem is in proportion to our enormous use of wood, to our unparalleled economic and social dependence upon forests. To solve this problem with even meager success, every resource of American ingenuity and foresight must be employed.

"We must find, almost overnight, a fresh source of raw material sufficient to supply sixty or seventy million tons of forest products annually. Instead of a gradual industrial evolution, the change is coming with the suddenness of an economic crisis."

Virgin timber was indeed being mined, and Greeley and other foresters were gravely concerned that the timber of the South and West would be as quickly depleted as that of the Lake States had been. He contrasted that with New England. "In many localities, and conspicuously in New England, timber production on private lands is already very profitable. The timber crop is proving the salvation of many a New England farm which has been pushed to the wall in agricultural competition."

Richard Fisher had a closer and more nuanced perspective on the situation in his native New England. He couldn't paint such a rosy picture and had his own missionary work to do. He wanted the box industry to survive, but he wished it to use a smaller percentage of the white pine resource. Fisher encouraged the industry to focus on the higher-quality products that might be sawn from the tree. Clear white pine makes great doors, window sash, floors, and trim around windows and doors inside and out. It can be made into furniture. But those uses require it to be grown to larger diameters. He wanted landowners to work on growing woods full of trees that were "better than box."

He encouraged them to cut only larger, older timber because "small timber means high costs and a low yield of better grades. Too much timber is being cut that is not only hard to market but would be making more money for its owner if left standing. The bigger the average log or tree, the lower the cost of lumber, and better the average quality. The place to start grading is in the woods before the trees are cut."

Richard Fisher was an English major before he went to forestry school. He was in the original cohort of foresters minted by the Yale School of Forestry, receiving his master's degree in 1903, and he had a reverence for both wild woods and well-wrought words. He found both in the work of his favorite writer, Henry David Thoreau. For those of us who know Thoreau only from *Walden* and his essay "Civil Disobedience"—or worse, maybe only from the aphorisms plucked from each—it may come as a surprise that he was a remarkably accomplished ecologist who saw and understood what generations of foresters have since been taught in school.

Thoreau read an address called "The Succession of Forest Trees" to the Middlesex Agricultural Society in Concord, Massachusetts, in September 1860. It shows his remarkable powers of observation and thoughtful analysis of what he saw. Of primary interest in our discussion here is his explanation of the oak-pine forest.

He described selecting for study a pine grove that seemingly contained no other species of trees. "Standing on the edge of this grove and looking through it, for it is quite level and free from underwood, for the most part bare, red-carpeted ground, you would have said that there was not a hardwood tree in it, young or old," Thoreau wrote. "But on looking carefully along over its floor I discovered, though it was not till my eye had got used to the search, that, alternating with thin ferns, and small blueberry bushes, there was, not merely here and there, but as often as every five feet and with a degree of regularity, a little oak, from three to twelve inches high, and in one place I found a green acorn dropped by the base of a pine."

He explained that the oaks came from squirrels, gray, red, and striped (chipmunks), and blue jays, which had buried acorns there for later. Those forgotten in the winter sprouted in the spring.

Later in his address, Thoreau said: "After seven or eight years, the hardwoods evidently find such a locality unfavorable to their growth, the pines being allowed to stand. But although these oaks almost invariably die if the pines are not cut down, it is probable that they do better for a few years under their shelter than they would anywhere else."

The regular seeding continues for years, with new acorns replacing withered oak saplings as long as mature oaks grow somewhere in the vicinity. When one year the loggers arrive and cut the pine, its removal floods the forest floor with a sudden flush of light, and the oaks, now free to grow, have all they need to take over the site. What was once and temporarily a pine forest develops into an oak forest. Thoreau said that if the cut occurred in a year that the pine seed was dropping from its cones, pine would seed in alongside the oak, and the two species would share the stand. If not, pine would disappear and the reversion to oak and other hardwoods would be complete.

Whether Fisher learned about forest succession from Thoreau, from his forestry professors at Yale, or from trial and error, he came to understand the dynamics of these old-field pine stands. In describing the silvicultural beginnings at Harvard Forest, he wrote, "It early became apparent that white pine in pure stands was a transition type, or intermediate stage in the succession, and one not in the long run suited to the comparatively heavy soils of the region. Always containing a variable admixture of hardwood, these forests develop in later life not an advance growth of pine but one of broadleaf species such as white ash, red oak, sugar maple, black and yellow birch, red maple, etc. This advance growth is not only a serious obstacle to the reproduction of the pine by any method of cutting but an index of a strong tendency toward reversion to the original or virgin type."

Pine was temporary, as all the foresters grudgingly learned. But it was the king of lumber, and many tried to prolong its reign. At that time, hardwoods like oak and maple were cut more for cordwood than for lumber. To be sure, there was a market for hardwood logs, but it accounted for only 20 percent of sales. The most desirable species at the time was white ash, the hardwood that perhaps best shares white pine's ability to grow quickly, thus putting on more board feet per year than the other hardwoods.

This obsession with pine might come as a surprise, since hardwood now has greater value. Today black cherry, sugar maple, and red oak are the most valuable species, followed by white ash, yellow birch, and paper birch. It's not that white pine is not highly valued. It is. But on a board foot basis, good hardwood logs are more valuable than pine logs. A pine sawmill might pay $350 per thousand board feet for white pine logs, while a hardwood mill will pay twice that for black cherry and $600 per thousand board feet for sugar maple.

White pine's beauty then and now is in growing so large so fast. Foresters tried to grow white pine wherever they could because it had proven to be such an asset. One way was through planting pine seedlings in abandoned fields, which accelerated what Thoreau saw as the process of succession. They attempted the same in cutover stands soon after the harvest, the site prepared by burning all the slash, which would otherwise impede the growth. Foresters encouraged farmers to plant white pine and red pine in thousands of acres of old fields, and following the lead of their European counterparts, they also introduced plantings of new species they hoped would prove reasonable surrogates for white pine: Norway spruce, white spruce, Scots pine, and European larch. The substitutes weren't up to the task. White spruce grows beautifully farther north in Vermont, but it grew poorly when planted in Massachusetts. Worse, almost all of the Scots pine looked more like a corkscrew than a pole, and remnants of those failed plantations can still be seen today.

Fisher planted his share of trees at Harvard Forest, but early on he developed an approach that worked with the natural system rather than against it. He saw that white pine suffered much less weevil damage when it grew among hardwoods, and he advocated growing mixed stands as a means to better pine.

Silviculture is the branch of forestry closest to agriculture. Its purpose is to grow a crop to be used or sold. Fisher's goal was to grow better pine and teach landowners to follow his lead. He wanted to convert people with the gospel of forestry. Fisher and the other evangelists knew that they had a crisis on their hands, that the industry couldn't get away with run-of-the-mill pine any longer. If white pine was to have value in these challenging markets, it had to be tended, and they knew how to do it. Trees with the worst

Richard Fisher demonstrating the technique of weeding the forest with a machete.
Notice that the saplings include hardwood species and white pine. He's removing
some of the hardwoods to favor the white pine, which was more valuable at the
time. P. R. Gast, 1925, Harvard Forest Archives, Petersham, Massachusetts

weevil damage should be removed as soon as possible, allowing the
better trees to absorb more of the site's sunlight, moisture, and nutrients. The trees with promising form should be pruned so that
the butt log—the first sixteen-foot log—would add plenty of clear
wood each year. And the trees should be grown to large size. Young
stands of mixed hardwoods and pines should be weeded so that the
hardwoods sheltered the pine from the weevil for long enough to
develop a good straight bole.

The Harvard silviculturists demonstrated these principles to landowners in their own forests, and Fisher even commissioned the creation of museum-quality forest models (the renowned dioramas now housed in the Fisher Museum at Harvard Forest) that displayed the techniques. They wrote bulletins and did everything in their power to proselytize.

Unfortunately, the gospel of forestry meant sacrifices. These labor-intensive treatments required a commitment from a landowner, and only those active managers who held forestland as a business paid any attention. Tending trees was a hard sell because landowners' recent experience with the pine seemed to prove that trees grew just fine and profitably on their own. Given the state of the economy, few people had faith that their investment of time was going to pay off.

Richard Fisher took pride in the white pine he tended at Harvard Forest. In a 1929 promotional piece, he described its holdings in this way: "Well timbered almost all over, it contains a greater variety of the different stages of forest represented in New England history than can be found on an equal area anywhere else in the region."

Fisher touted the outreach accomplished by the forest's staff in demonstrating how to make money in this pursuit. "One of the principal reasons why the lessons of the Harvard Forest have been convincing is that it has been managed from the start as a paying enterprise in which the developing measures for maintaining and increasing production were more than paid for by the income from cuttings. Today, the Forest has a larger volume of timber and a substantially higher annual production than it had twenty years ago; and in the meanwhile nearly four and one-half million feet of lumber have been cut and profitably marketed."

By weeding and thinning and selling the weeds, Harvard Forest staff had greatly improved the quality of its timber. After offering sterling testimonials from leaders in the boxboard and paper industry and various government officials, Fisher came to the unexpected point: "At present, although the Forest is producing as great a revenue as ever, there is an annual deficit, which has been met by annual gifts. If the fruits of twenty years of building are to be realized and secured for the future, the Forest must have a substantial increase in its resources."

Fisher's fund-raising appeal makes abundantly clear how challenging it is to make a living from a piece of forestland. For Harvard Forest and for thousands of landowners in New England, that challenge was soon multiplied as the demand for timber was crushed by the Great Depression. In 1929, a total of 15.7 billion board feet of timber was cut in the entire United States. The average cut in the next five years was a quarter of that, 4.1 billion board feet.

New England's white pine industry had already contracted in the 1920s, and then consumers stopped buying lumber in the 1930s. Instead of sawing logs, portable mills were swallowed up by encroaching trees. Even worse than its neighbors to the north, Massachusetts lost much of its wood products infrastructure in the 1930s. The boom had officially gone bust. Landowners paid little attention to their trees because nobody was buying them.

Given a reprieve for a decade or more, pine trees added inches in diameter and feet in height. The U.S. Forest Service conducted a national inventory in 1938 and reported that the Lake States industry had cut more than 99 percent of its virgin timber. To show how far the white pine pendulum had swung back east, it was estimated that of the 18 billion board feet of white pine in the nation, 10 billion of it was in New England. The highest inventory of white pine in decades stood tall in central New England as Thirty-Eight came roaring through.

FIVE

A Day at the Fair

When I first visited with Jim Colby in 2012, he lived just a mile down the road from the farm where he grew up, in Boscawen, New Hampshire. Earlier that week, Colby, ninety-two, had buried his ninety-five-year-old brother. Time had been kind to Jim, and maybe it was because he didn't avoid it—each room in his house murmured with old clocks that tick every second and chime every quarter hour. Several clocks gave him some trouble, so the announcement of the quarter hour chimed on for a few minutes.

In his blue work pants and blue flannel shirt, he looked ready to go to work at the sawmill. He and his family—starting with his father and now his sons—have been in the lumber business ever since Thirty-Eight blew through and littered the ground with timber. However, on the day the rivers swelled and the wind blew, it wasn't trees he was concerned about, it was livestock.

His father, Joe Colby, was a judge at the nearby Hopkinton Fair, and even though Wednesday was a special day at the fair, it had rained so hard they closed it down. His father came home and picked up Jim, and they crossed the bridge over the Merrimack to help a farmer friend, Ralph Graham, in Canterbury. Graham had seventy-five or eighty purebred Holstein cows, and the water was coming up and would swamp the cows if they stayed where they were penned. "We got three of them across the bridge to the barn, but we couldn't get the rest of them. They just wouldn't go," Colby said.

"When we were trying to rope those cows over, I'll never forget it, here comes a ten-by-twelve little henhouse floating down the river, and there's eight or ten hens and a rooster sitting up top, and the rooster was just crowing away."

As the river kept rising, they could see that they were quickly running out of options for the rest of the Holsteins. Finally, Graham saw what they should do. "The house was up higher, so we put the cows in his house," Colby said. "It spoiled the house. It was a nice old house and he had to go through it all after, fixing it back up, but it saved the cows."

Then Jim and his father went to help another farmer, Scott Sanborn, who had both sheep and cattle. Colby said, "We had to get the sheep up where the water wouldn't get them, so we put them in the barn up on the high beams, up in the haymow, and boarded them off. But he also had a registered Guernsey bull over there, and he said, 'I can't lose that bull.' He'd paid two hundred dollars for him, and back then two hundred dollars was like ten thousand."

The only way to save the young bull was to bring him across the Merrimack from Canterbury to Boscawen, and the only way to do that was to put him in a boat. "Why my father ever wanted to go over there and why I went with him I'll never know, because I couldn't swim. If I did, I'd have sunk anyway. So would my father. John Stone was a friend of my father's, and he had a good-sized rowboat. He was a good man, a rugged man, and so was my dad. They got that bull, tied his feet together, tied the back ones right up to the front so he couldn't do anything, and put him in that damn boat and away we went."

The current was heavy, and Stone knew that he had to get them upstream far enough that when they set out across the current, it wouldn't take them too far downstream on the other side. "He rowed the boat as much as he could upstream, and then we shot right across into the current, and he was a pumping on one oar and he got it to shore under the Canterbury Bridge, where we were going to land. I'll tell you, if there's a God in heaven, He was there that day. I wouldn't have gone across again if he had paid ten thousand dollars for that bull."

Their livestock rescue duties behind them, the Colbys went back home, just in time to see all the trees in their dooryard tip over. "There were elms and locusts, and somebody must have set them out because some of those elms were two and a half or three foot across. You could just see those trees tip over one by one. The wind didn't really hit until about six o'clock. When it came, it was just like a siren." Jim pronounced the word like many New Englanders do: *sireeeen*. "It came in *sssss*, and then it was over. I don't think it lasted more than three-quarters of an hour, or an hour." Then the wind stopped screaming and left behind an unearthly calm. Colby said, "The only way I can explain it to you is it seemed like we were living in a vacuum after that wind went down. It just seemed like you were in a hollow cave."

The Colbys and the Grahams were farming in the Merrimack Valley in 1938, but they were two of the few. The 1940 census for Boscawen shows that of 389 heads of households, only 7 percent were farmers. The bulk of the others worked in woolen mills or other factories. Contrast that with Corinth, a typical town in central Vermont, where 45 percent of the heads of household in 1940 were still farming.

The Hopkinton Fair, where Joe Colby was supposed to be judg-
ing cattle, wasn't the only fair to be shut down that day. Earlier, starting at around four o'clock that afternoon, the biggest fairgrounds in
New England had been ripped apart. Visitors from all over New
England had traveled to Springfield, Massachusetts, to attend the
Eastern States Exposition. What is now known as the Big E has long
been the biggest combined state fair anywhere. Each of the six New
England states has its own building filled with livestock, produce,
equipment, and educational exhibits from extension agencies. Fairgoers spend the day meandering through their own state's buildings, and occasionally a Vermonter might admit that he even took
a peek at what was going on in the New Hampshire or Connecticut building. The highlight of the 1938 Exposition was the newly
constructed Grange Building, which had its dedication ceremony
that morning. The dignitaries made their speeches, the people
clapped, but by then, the soggy fairgrounds was already a mudhole,
prompting the early departure of many of the fairgoers, including
Wentworth Blodgett, of Bradford, Vermont. If he hadn't left on his
own, Blodgett would have been part of the throng that was evacu-
ated from the fair when the winds came and took the roof off of one
building. The Grange Building held up fine, but the Ferris wheels
blew over, as did countless tents.

One hundred thirty miles upstream of the carnage in Springfield,
Blodgett's wife and three children had no idea what was coming.
Perched on a shelf above the Connecticut River in Bradford, Vermont, the Blodgett farm provided a beautiful view of the wide, lush
valley, known for some of the most productive agriculture land anywhere. In 1938 he owned four hundred acres of land, including
more than three hundred acres of woods dominated by white pine,
where cows were allowed to graze.

The hurricane was moving forward at 47 miles per hour, so it
took nearly three hours for the storm to bring its punishing winds

The Eastern States Exposition in Springfield, Massachusetts, began on Sunday, September 18, 1938, and officials made a valiant effort to keep the fair open despite the days of rain that turned the grounds into mire. Then the hurricane ripped through, and for the first time ever, the fair was closed down early. This photo of the east end of the fairgrounds shows what's left of the exhibit tents and animal sheds. In the background, you can see three Ferris wheels twisted together. Courtesy of Eastern States Exposition

to Bradford. The storm's trip north was more direct than Blodgett's. His son Putnam (known as Put) was seven years old at the time, and he recalls his father telling of the hard time he had getting home. Put said his father had to go up the New Hampshire side beyond Bradford because he couldn't get across on any of the bridges downstream. Then he couldn't drive on the road along the river because of the high water. "Cow Meadow was flooded, as it always does between Wells River and Newbury, so he had to wind his way down through the back roads of West Newbury and Bradford to get home."

When Put's father arrived after dark to find the barn doors flapping in the wind, he tried to prop them shut. No matter. One of them went cartwheeling across the yard. He learned that his family had already had quite a fright. Put said, "We kids had been fed and put to bed, but we had two giant elms in the yard that my mother was afraid were going to come through the roof so she took my brother and sister and myself downstairs. The electricity was out and she had prepared a cold supper for my father and the hired man, and she had it all laid out on the dining room table. Then a limb from one of the elms came through the window and scattered broken glass all over the meal."

Meanwhile, in the Vermont hills fifteen miles to the west, Harold Luce, a twenty-year-old fiddler from Chelsea, Vermont, had played for the contra dances at the Tunbridge World's Fair that afternoon, just as he had the previous four years. This little country fair has been known as the "world's fair" since 1867, and it features harness racing, ox and horse pulls, cotton candy, displays of blacksmithing, timber hewing, shingle making, spinning, weaving, contra and square dancing, and fiddling contests. Like most fairs, it has a Ferris wheel and merry-go-round and a midway where carnies help you part with your money. These fairs have always coincided with the late-summer harvest, and piemakers, gardeners, and canners bring the fruits of their efforts to compete for ribbons. Youngsters in 4-H show their steers and calves, lambs and kids, chickens and rabbits. A celebration of farm life and all its traditions, it's also an annual opportunity for farmers to get off the farm and have a good time not working for a change.

"I had played that day," Luce said, "but they quit a little bit early because it got cloudy and got so dark they thought everybody better go home, so we went home. The wind did a job on the tents down at the fairground. Blew most all of them over and ripped 'em apart."

Luce remembers that night's supper. His father's customary place

Fiddler Harold Luce performs for the contra dance at the Tunbridge World's Fair in 1939, the year after the hurricane blew the fairgrounds apart. Performing with Luce are caller Ed Larkin and Walter Sawyer on the melodion. Courtesy of the Tunbridge Historical Society

to sit for meals was in a chair at the back side of the supper table. That evening, his father said he could feel the wall moving in and out with the gusts of wind, and he was worried the house might go. It held on, but they lost a swath of trees.

"That was the height of the storm," Luce said. "I then had to go out in the dark and get the cows, and I didn't think I was going to get home. They were down in the pasture on the south end of the woods. They had gone up into the woods to get protection. I had to hunt for them before I found them, but I got them home."

"Tell me what the wind was like," I asked.

"I can't tell you. Terrible," he said, and didn't elaborate.

"Were you scared?"

"Yes."

■

Between Put Blodgett's farm in Bradford and Harold Luce's farm in Chelsea sits the town of Corinth, the town where I've lived since 1988. The debate over the pronunciation of the town's name has gone on—*raged* would be too strong a verb, but *smoldered* might work—since its incorporation. Some say COR-inth, others say Co-RINTH, rising in pitch and relegating the *o* to a mere placeholder. In fact, the closer you go to turning it into a single-syllable word, the better. I hold with the latter, though given my lack of standing as such a recent arrival, it doesn't much matter what I think. Old-timers seem split on the matter.

As for many towns in Vermont, controversy also attended its origins. Both New York and New Hampshire claimed the land that lay between Lake Champlain and the Connecticut River, so the town has competing charters and lotting surveys. The current boundaries are those that were set out in the New Hampshire charter, but in my section of town, many of the deeds have as one of their bounds what is referred to as the York Line. The original deed for my land had it bounded on the north by the York Line, which would have put us in Vershire, the town to the south.

The various hubs of commerce—Montpelier, Saint Johnsbury, Randolph, and Norwich, with its across-the-river sister, Hanover, New Hampshire—are each a forty-five-minute drive. That means something different today than it meant in 1938. Today, it means that most people have a forty-five-minute commute to jobs. In 1938 it meant that most commerce was conducted locally. The town supported four stores, and nearly half of the people relied on farming for their living. Except for the milk, which was shipped out of town, most of the farm's production was used on the farm or sold or bartered to neighbors.

The oldest section of town, Taplin Hill, was settled in the late 1780s, five decades later than the towns of Rindge, Boscawen, and Petersham. In the same general pattern that held across New England, long broad hilltops were settled before the narrow river valleys because people found them easier to clear for farming. Only later would settlers need a commercial center with water power, and East Corinth, built down below on the banks of a branch of the Waits River, soon provided that to the residents of Taplin Hill. Many of the people scouting out what would become Orange County in Vermont were the children or grandchildren of those who'd settled Connecticut. Those of the younger generation decided to strike out on their own, and many of them liked the farming prospects on top, because the upland forests featured sugar maple, white ash, and basswood, a collection of species that they knew grew on good loamy soils. In the deep shade beneath the canopy, they found maidenhair fern and ginseng.

The soils are remarkably productive, benefiting from the local phenomenon of calcium-rich bedrock. The challenge to Corinth farmers has been not the fertility but the topography. Crumple up a sheet of paper and do a cursory job smoothing it out and you'll have an approximation of Corinth's topography. If you happen upon any flat ground, it is more likely to have come courtesy of heavy equipment than from natural forces. The pioneering farmers cleared the forests except for the steeper ridges, which they kept as sugar orchards, and mixed wood forests in the wetter depressions that provided hardwood cordwood and lumber from spruce, fir, and hemlock. They grew corn, wheat, and oats on the gentle lands, cut hay on the sidehills, and pastured cows in the cleared land that couldn't be hayed because it was too steep, too wet, or too rocky.

They built up their farms, and these soils supported generation after generation of farmers. Nearly 140 years after settlement, in the midst of the Depression, eight family farms were still making a go of it on Taplin Hill. They tended small herds of cows—twenty to forty milkers was typical in the days before every farm had to have its own bulk tank—and shipped milk via the milk truck that came by every day. Each family relied on a spring crop of maple sugar to supplement its income.

Bryce Metcalf, Harry Brainerd, and Dustin White were all twelve years old when the hurricane hit. Lois Worthley was a year older, and these four scholars—as they were invariably referred to in town reports of the time—made up one-third of the entire enrollment in the one-room schoolhouse on Taplin Hill. Each of them remembers the storm well.

The Brainerd farm is one of the oldest in town. The white Cape-style house and large white barn perch on a knoll with an iconic sugar maple in the yard, today providing one of the nicest photographs in town. In 1938 they provided a nice target for a fierce wind. Harry Brainerd was suffering from rheumatic fever at the time, so he had been carried to and from school that day. "Both of my knees were swollen so, and when I came home in the afternoon, I was sitting by the window. School didn't let out until 4 o'clock back in those days, and I was sitting in the chair right there. I noticed that the little trees were bending right over about twenty minutes past four. I thought 'twas awful windy, but we just didn't pay much attention." (The rest of New England was on Eastern Daylight Time, but Vermont adhered to Standard Time, so Brainerd's time estimate would equate to 5:00 and 5:20 elsewhere in New England.)

As it got progressively darker, the wind kicked into another gear. A mile or so to the west, Bryce Metcalf was looking south at the

Hood farm across the road, and he could see the maple trees start toppling, roots and everything. "We knew then that we had something to contend with," he said.

His father was worried about the roof on their barn and decided to try to clinch the nails that were holding the metal roof to the roofing boards inside the barn. By bending the protruding nails over, he hoped to secure the roofing. "The barn was full of hay, loose hay," Metcalf said. "We used a hay fork to pitch it right up to the roof, so we got right up on top of that hay. I was twelve years old, so I was trying to help my father and the hired man. The lights had gone out by then, so we had a lantern up there, we were clinching the nails, and then all of a sudden my father could see the whole roof starting to lift up. He said, 'We'll get out of here.'"

Lois Worthley, now Lois Sherwood, said, "It was getting on toward night and the wind was pretty serious, and my dad wouldn't let me go to the barn to do my chores. He didn't want me to blow away, I guess. We had standing-seam roofing, and the wind got under that so it was clattering. I remember that noise. My dad had a hard time getting from the house to the barn. The barn sets northwest to southeast, and they had built a shed perpendicular to it. He kept his farm equipment in it, and it was open on the east side. That shed blew over. The wind just picked it up and dumped it.

"It also blew over most of the older maple trees in the woods. Ripped 'em right up so that the whole root system was standing up."

Back at Brainerd's, more of the same. Harry told me,

We had a Wards airline radio and the aerial was across on the old barn. That came flying down and we decided that the wind was really blowing. We used to listen to Adolf Hitler and the Nazis about three nights a week, and they had an interpreter that came on and translated into English. We thought, well, with the aerial down we won't be able to listen to those birds.

We used to listen to Lowell Thomas as our regular American commentator. Of course, that was the beginning of news on the radio.

We ate supper and the wind just kept getting worse. Then we looked out and our barn doors were going out across the field end over end. It took the doors right off the big barn. About then the first tree went and came right on the doorstep. My mother had just come in, she'd got the washing in from the clothesline, and then the pine and the spruce started hitting the house, and when they hit the ridge pole, they broke right off, and the top part of the tree slid down the other side. It just got so that every time one of them huge trees struck the house, the house just jumped right up and down. I remember my father said, and my mother too, "I guess we'll be starting over again." We figured the barns would probably be demolished and everything. That's the way it was.

The White farm faced the brunt of the storm as it stood on the south flank of Taplin Hill, and Dustin White to this day marvels at the fact that they didn't recognize that the storm was anything out of the ordinary. "When we went to bed, we knew the wind was howling bad, but we thought it was just a bad wind. I remember waking up in the night and the wind was howling and the house would shake. We knew we had a heck of a wind, but nobody knew we were in the midst of a hurricane."

White continued, "When we got up the next morning, there were trees down all around. It blew down the old horse barn. We kept hay in it for the horses and it flattened that to the ground, but it didn't flatten the building the horses were in. That was kind of hitched to it, but that stayed there, while it flattened the hay barn.

"Dad had a five-acre piece of field corn and he hadn't gotten around to cut it, because he was waiting for the corn to dry down.

When we got out and got around, we went up the hill, and that corn was just like a steamroller had run over it. The whole piece was just flat to the ground. There wasn't any corn left standing."

Other farmers recall similar scenes of flattened corn. Harold Luce used the same exact words: "It was just like a steamroller had run over it." The frugal Yankees managed to harvest their cornfields by hand. White said, "The main thing they thought about was salvaging that corn piece. That took them a while. Five acres, picking up every stalk. I remember them commenting what a nice fall we had after that; the weather was nice, it stayed warm. They were able to get all that corn by picking it all back up off the ground and taking the ears off and then they cut the corn and kind of stook it."

Stook rhymes with *kook* and means to assemble the stalks into a tepee shape so the corn could dry. White's routine use of words like *stook* harks back to his grandfather's time. His accent features a cadence and inflection that has nearly disappeared from this section of Vermont. When he pronounces the word *hurricane*, it sounds like HAIR-a-cane, making it seem just as exotic as it did to all those people who in 1938 had their first taste of one.

Their immediate concern might have been to salvage the steam-rolled corn, but the longer-term challenge was in their sugar orchard, which had been broken apart. All the local farmers tapped their maples to make syrup and sugar from the sap. The Whites would set out twelve hundred buckets each spring and as many as sixteen hundred if they got an early enough start. Lois's father, Irwin Worthley, had one thousand taps and the Metcalfs sixteen hundred on their forty-acre sugarbush, known in the local vernacular as a sugar place. The Brainerds had their sugar place, but adjacent to them was a landmark among sugarmakers, the Maple Farm, owned by Elmer Eames.

Harry Brainerd said of the Eames place, "It was beautiful, the best piece of hardwood around. Big nice trees, they went up there straight as an arrow. It was the prime sugar place in the town of Corinth. It won the grand prize blue ribbon at the 1893 International Exposition in Chicago. They used to hang two or three thousand buckets out there and had two ox teams gathering.

"Out here in the dooryard, you could hear their trees falling. It was just awful. Of course, everybody's sugar place went down. When that wind got hold of the tops of those trees, the leverage on it pulled them right out by the roots and tipped them over. It was a catastrophe. Afterward at the Eames place, they could hang just six hundred buckets." Twenty-five years later, Brainerd would purchase that land and wait patiently for it to develop once again into a world-class sugarbush.

The annual rite of spring was a crucial part of the farm year in Vermont. Lois (Worthley) Sherwood said, "We always made granulated maple sugar for our own use. My dad got a new big square pan they used to mix cement in, he got a new one and a couple of new hoes. My mother would boil it down to the right degree, and we would get on either side and work it back and forth with the hoes. Then we put it up in buckets, and that was basically all the sugar she had. Later, when they had sugar rationing, she didn't care." She always had all she needed.

Perhaps more important, sugar was a cash crop they could make at a time when it was otherwise slow on the farm. Harry Brainerd said, "Everybody look forward to sugaring in my day because it gave you a little bit of extra money to buy things, clothes or things for the kids. It gave you some extra money to do things with. It didn't matter how deep the snow was nor how hard it was to get the sap."

Harry's wife, Joan, grew up across the valley from Taplin Hill. She said, "Everybody made maple sugar. I don't know as they were really contests, but every family strived to see who could make the whitest sugar and the clearest syrup, so you had a natural competition. Irwin Worthley was very particular with his sugar cakes. I don't

The Worthleys and other sugarmakers shipped maple sugar in boxes like this one, which was made at a box shop in East Corinth, Vermont. The wood, as usual, is white pine. Stephen Long

think anybody could match him, really. When his sugar cakes went for sale at the store, they were just as square as could be and there wasn't even a corner that showed any sign of a crack. It had to be perfect."

Worthley's daughter Lois remembers it the same way. "We had a really good orchard, and my dad said it was because it had a north exposure. It was very light, and as a matter of fact, they were often accused of putting white sugar in it."

The cumulative effect of all of these farmers' reliance on sugaring is documented in the Statistical Abstract of the United States for 1938. It showed that Vermont's sugarmakers hung 5.4 million buckets that spring compared with only 368,000 in New Hampshire and 224,000 in Massachusetts. With its 2.9 million buckets, New York was the only state to come close to Vermont.

Each Vermont tap produced about 2.3 pounds of sugar, a level of output that squares well with the rule of thumb that existed before sugarmakers went high-tech and introduced pipelines, vacuum pumps, and reverse osmosis. In the days of buckets, sugarmakers expected each tap to produce eight to ten gallons of sap over the course of the season; boiled down, that would produce one quart of syrup. With sixteen hundred buckets, the Whites and the Metcalfs could expect an annual crop of four hundred gallons of syrup, or, if they chose to boil it down further, thirty-two hundred pounds of sugar. The 1938 abstract reports that 95 percent of the maple production was in the form of syrup, and a Vermont Department of Agriculture bulletin from 1930 gave the percentage of syrup as 84 percent. The Taplin Hill sugarmakers remember it differently; independently of each other, they told me that the bulk of what they made was sugar cakes.

Dairy farmers were accustomed to selling their milk wholesale to the cooperative or to a distributor and not directly to consumers. Syrup and sugar are not perishable like milk, so the sugarmakers didn't need to unload their product right away. That meant they could sell it retail, and the Eames Farm in particular did a lot of mail-order business. Already, the nostalgic angle of pure, wholesome Vermont had found its way into the marketing of maple syrup.

Bryce Metcalf said that his family made cream sugar, syrup, and lots of sugar cakes: "You'd put the sugar cakes in thirty-two-pound boxes and ship it to Boston." The boxes were, of course, made of white pine, and they were driven to the railroad station at Bradford, about ten miles away.

Harry Brainerd recalls riding with his father down to Bradford with their sugar. "They used to ship twenty-five hundred to three thousand pounds a day out of the old Bradford railroad station. Most of it went to Boston or to other cities down country. I could see the boxes down there piled this high."

A gallon of maple syrup is the equivalent of eight pounds of sugar cakes. In 1938 the Taplin Hill farmers could sell a gallon of syrup for about $2.00. Sugar cakes required more processing and brought them 27¢ per pound. If they produced four hundred gallons of syrup, they earned $800. The equivalent in sugar—thirty-two hundred pounds—would bring them $864. This was more than pin money.

The 1940 census includes income figures for employees who were paid a wage, but not for those who earned their income from a business like farming. To put that $800 into context, a farmhand on the Brainerd farm made $780 for the year, which included room and board. Leo Hutchinson paid his hired man less, $600 for the year. Both men worked fifty-two weeks. The rural mail carrier made $1,700, and a public school teacher made $665. Laborers at the bobbin mill in East Corinth ranged from $690 to $1,200. The local minister made $1,300.

For a month or two of work in the spring, a family could supplement its annual income with the equivalent of a full year's salary for a laborer. No matter the amount of work entailed, it was an essential source of cash flow.

The hurricane took out a third of the trees that the Whites tapped. The Metcalfs and the Worthleys lost even more. Lois Sherwood said, "That was a dire consequence of the hurricane: we lost the maple sugar orchard for probably five years. That was our cash crop. It supplemented the income not just during the season, but after as well, because people placed orders after."

Bryce Metcalf said, "We had four hundred acres of land, and maybe forty acres was sugar maples. We had some other woods, softwoods mixed with hardwood, and they got blown down some." But the next morning they could see how devastating the loss was to the forty acres of sugar orchard. "It was just a jungle, you couldn't even walk out there because so many trees were down," Metcalf said. "A good half of the trees went down, and the trees that were left standing had branches and limbs blown off so they never ran sap like they did previously. It pretty near ruined it."

The Bowen-Hunter Bobbin Mill in East Corinth made bobbins for the textile industry out of sugar maple logs. It purchased much

In Thetford, Vermont, the maple trees and the sugarhouse on this east-facing hillside were exposed to the winds from the southeast. United States Forest Service

of the downed hardwood in Corinth, including the remains of the Metcalfs' sugar orchard. Sawmillers are careful even today about sawing maples that have been tapped for fear that there might be some spouts hiding in the wood that will break the saw's teeth. Metcalf said, "When you tap, you use metal spouts, so they had to butt all the logs. They'd been tapping these trees maybe fifty or a hundred years, and they might have left some spouts in there."

That meant the loggers left the bottom six feet of each maple tree where it fell. The butt log is the largest and most valuable part of the tree, so leaving them behind greatly reduced the volume and value of what was salvaged, but still, they cut fifty-two thousand board feet of timber from downed maples. The Metcalfs were paid about ten dollars per thousand board feet, so they recouped $520, not much recompense for the goose that would have continued laying golden eggs for decades to come.

■

For centuries, writers have been expounding on the differences between Vermont and New Hampshire, some of it scholarly, some of it humorous. Each state's partisans generally start with pulling out a map and demonstrating that the other state is situated upside down. The different topography, mountains, bedrock, and soils all point to the twin states' origins not from the same womb but indeed as parts of different continents, a startling realization brought to us through the study of plate tectonics. These differences also ensured a different history of land use. In shorthand, Vermont's richer soils favored agriculture, while New Hampshire's steeper stream gradients facilitated more water power and thus industry.

Given that Thirty-Eight's track went right up through the Green Mountains, eastern Vermont suffered through some of the storm's most damaging winds. Even though it was farther from the storm center, much of New Hampshire experienced the same intensity of winds because the extratropical nature of the storm had extended the radius of its damage. A billion board feet of timber was blown down in New Hampshire, but only a third of that in Vermont. Part of that tremendous disparity came because the storm track essentially bisected Vermont, so only its eastern half experienced the worst of the wind. Another reason was simply that Vermont had fewer acres covered in mature forests.

The pastoral combination of farm field and forest has adorned Vermont's tourism publications for more than a century. Vantage points to take those iconic photographs can still be found today, even though forests have relentlessly invaded fields, leaving only 20 percent of the state unforested. Back in the 1930s, that combination of farm and forest could be found almost anywhere.

A thorough timber inventory conducted by the Forestry Division in 1926 showed that only 55 percent of Vermont held forests made up of trees greater than three inches in diameter. That's not a particularly high threshold to meet; it's akin to saying that someone with three days of stubble on his chin has a beard. So 45 percent of Vermont was open, and a good portion of the forest wasn't mature enough to blow down. The species mix was vastly different from New Hampshire's. Only in the Connecticut River valley, as at the Blodgett farm in Bradford, was white pine a substantial player. The 1926 inventory showed that only 3 percent of the forest was in pure stands of white pine. In another 7 percent, pine was part of the mix with hardwoods. Then and now, hardwoods dominate Vermont, much of it in pure stands, though some of it is mixed with hemlock, spruce, and fir. In the 1930s, 53 percent of Vermont's forests were in pure stands of hardwoods.

A smaller volume was lost in the Green Mountain State overall, but the relative importance of that loss to the owners can be seen as much greater. Instead of the white pine bank account that had been accruing interest for its owners only on paper, the Vermont sugarbush was a crucial source of annual cash for the farmers who still occupied these hill farms.

This picture of a strong agricultural presence in 1930s Vermont can be difficult to reconcile with the conventional wisdom of the abandoned rural countryside. The latter view comes from a number of sources, perhaps most notably from Harold F. Wilson's *Hill Country of Northern New England*, which told of the sorry fate of hill farms in northern New England, and particularly in his native state of Vermont. Writing in 1936, Wilson quotes from an 1895 publication, "A steady emigration takes away the young, the hopeful, the ambitious. There remain behind the super-annuated, the feeble, the dull, the stagnant rich who will risk nothing, the ne'er-do-wells who have nothing to risk."

Subsequent scholars have often cited Wilson's work, which showed that 1880 was the zenith for the number of farms in Vermont, New Hampshire, and Maine, and that the numbers declined thereafter. Emigrants left in droves, first to the West in the 1850s and 1860s, followed by migration to closer job opportunities in the increasingly industrial southern New England starting in the 1880s. As the young men and women went off to find work, the population on the farms aged. Rural towns steadily lost population, though the states continued to grow.

Wilson's scholarship is augmented by personal anecdotes from his family and friends, and he sees the last three decades of the nineteenth century as a severe winter season: "The shock of the widespread desertion of farms and the pronounced decline in rural population, with their social and economic consequences, stunned the hill country."

As in southern New England, it was the submarginal farmland that was allowed to grow back to trees. Again, the term *abandonment* might connote more than it should; rarely would someone throw up his hands and ride out of town. Instead, farmers who wanted to leave sold their poorly performing farms to other farmers, to loggers, or to speculators.

Vermont's population growth curve differs broadly from those of its neighbors. It was settled half a century later, and like a high performance car, it sped from zero to 300,000 in six decades, a remarkable pace for settlement of a wild land. Like its neighbors, it began to lose farmers in the 1850s, but that loss wasn't minimized by the growing urban industrial centers as it was in Massachusetts and New Hampshire. In Vermont, the growth of urban centers barely countered the loss in the rural counties, so its population flatlined for more than a century while the other New England states increased steadily.

Lost in the dreariness of Wilson's dirge are his various attempts to temper his bleak view. "Although the discussion of the innumerable factors causing the abandonment of farms and the decline of population may have given such an impression, the New England hill country had by no means become a wasteland of deserted homesteads and decaying villages. Thousands of fertile and promising farms covered the wider valley stretches and the upland regions where the land was rolling rather than mountainous, and hundreds of quiet little villages continued to raise the white spires of their churches above the surrounding trees."

Wilson also points out that the more machinery in place on the farm, the fewer people it took to raise the same amount of milk. And he acknowledges that in the hill country, farming was a way of life and not merely a way of making a living. Consequently, some of the hill farms were not abandoned and some probably would never be. But his overall picture of a stagnating region became the prevailing view.

Then along came Hal Barron fifty years later with a book called *Those Who Stayed Behind;* Barron looks at the same period and even the same data but tells a different story of Vermont in the late nineteenth century. Believing that Wilson painted with too broad a brush, he instead makes his case through a tight focus on the town of Chelsea, home of fiddler Harold Luce and one town west of the folks on Corinth's Taplin Hill. He examined census records and followed individuals from census to census to ascertain who stayed and who left. He found diaries of local farmers, farming publications, records of agricultural societies, and many other local sources to inform his understanding of the economy and community of this town, whose population crested in 1840 at 1,959 and dropped to 1,070 by 1900.

Barron shows that the out-migration of young people reduced the population but didn't reduce the productivity of Chelsea's farms. Once in place, many of these hill farms lasted for generations, and their continued operation contributed to stability in the community. What previous historians have referred to as stagnation and decline Barron sees as the stability of a frontier that has matured into an agrarian society. "The stability of Chelsea's agriculture over most of the second half of the nineteenth century, and the lack of abandonment in the wake of falling wool prices, belie the images of long-term economic decline and deserted farms that characterize the traditional view of rural Vermont during these years. Instead, the constant number of farms and level of farm production suggest that agricultural development in Chelsea had reached the limits of its growth and had leveled off."

Maintaining enough of a landbase for successful farming meant not dividing up the land among the children. Instead, it was generally deeded as a whole to a male child, often with the stipulation that the parents could maintain residence on the farm for their entire lives. Other children were provided for separately if there was any other wealth to spread around, but that was not a given. The European tradition of primogeniture, in which the land went to the firstborn, did not continue in the colonies. There were no hereditary titles to be passed down with the land, and nearly everyone owned land, so the line of descent was meaningless. And even more practically, the eldest son of a settler who took to heart the dictum to be fruitful and multiply might be only twenty years younger than his father. It might take a long time before the eldest could wrest control of the farm from Dad. Instead, older sons generally left the farm. Some bought their own land nearby or learned a trade in the village; others went west. Most often, the farm was eventually transferred to the youngest son, who would not be competing with his father for control of the operation until later.

Young people left, birthrates of the aging population decreased, and few new people sought their fortune in staid established towns like Chelsea. But the amount of land used for farming stayed relatively constant once the least productive land was abandoned. In each census from 1840 to 1900, Chelsea always had about two hundred farms. Barron noted that Chelsea typified towns in Orange County in its economy and community. He writes, "Those who stayed behind in older rural America were not necessarily left there by the progressive advance of an urban and industrial society. Many chose to stay and experienced a sense of stability that eluded a good number of their contemporaries."

Farms clung to sidehills well into the 1940s. In Corinth, Dustin White recalled bringing his future wife, Jane, for a visit to the farm on Taplin Hill. She had grown up on a hill farm in East Montpelier, so she was familiar with the terrain. White told me, "Jane wanted to know 'how come anybody would farm on these steep places.' Well you stop and realize it, everybody had a farm. I said we couldn't all have good laying land, we had to take what we had."

Across New England, you can roam through the woods and find cellar holes of long-gone barns and farmhouses that were overtaken by trees long enough ago that the trees are now mature. If it were possible to ascertain the date of abandonment, chances are we'd find that those in Vermont held on longer than in New Hampshire or southern New England. In the 1920s, Vermont was the only state that had more cows than people. When Thirty-Eight blew into Vermont, it didn't find the easy pickings of old-field white pine as it did elsewhere. Instead, it steamrolled corn and blew off barn doors, but in its quest for trees it had to seek out the mature forests on the ridges, places that hadn't worked for raising cattle but had suited the sugarmaker very nicely.

■

It would be convenient to reduce the hurricane story to the difference between sugar orchards in the north and old-field white pine elsewhere. But if that were true, how would we explain what happened in Connecticut, where hardwood made up such a significant portion of the loss? In any place at any time, the current forest cover springs from the particular long-term interaction between human history and natural processes. In Connecticut, the extent of the hurricane damage was colored by its markedly different human history and a recent ecological disruption.

Some sections of the state with particularly fertile soils maintained an agricultural base akin to Vermont's. Connecticut had also been one of the nation's earliest centers of industry, mining ore from its rock and processing it with its trees. One focus for a young vulnerable nation was on arming itself. Eli Whitney, who is more widely known as the inventor of the cotton gin, started the firearms industry in Connecticut in 1798 when he made thousands of muskets for the army by using interchangeable parts. In the century that followed, all of the big names in guns—Colt, Browning, Smith and Wesson, Winchester, Remington, Savage, Gatling, Sharps, Marlin —had factories in Connecticut.

Stanley Tools has made hardware and hand tools in Connecticut since 1843. Pratt and Whitney, today building jet engines, started out with milling machines in 1865. All of these companies and thousands of others used iron and other metals, whose processing required enormous amounts of charcoal. A collier would build a solid dome of small hardwood logs (between one and seven inches in diameter, cut into four-foot lengths) around a central wooden chimney. He would stack it tightly fourteen feet tall and thirty feet in diameter, making it airtight by topping it with leaves, ferns, and dirt. He'd tend the pile for as long as fourteen days while everything was burned out of the wood but the carbon. A cord of hickory, oak, and beech smoldered down into thirty to forty bushels of charcoal. Colliers used wood from thousands and thousands of acres of these hardwoods, clear-cutting and returning in thirty years when the stump sprouts had reached burnable size again.

The most prolific of the sprouters was American chestnut, which easily outpaced the other species and was therefore grown to larger sizes because of its value for higher uses. A recitation of its qualities suggests that it was indeed a perfect tree: fast-growing, with wood that was strong but relatively light, and even rot-resistant. It also produced prolific crops of nuts prized by animals and humans. It could be used for split rail fences, railroad ties, telephone poles, shingles and clapboards, and furniture.

Chestnut had been a relatively minor component of New England's presettlement forest, but by 1900, it dominated Connecticut forests to the same extent that white pine dominated Massachusetts. Austin Hawes, Connecticut's state forester, wrote in 1906 that "chestnut constitutes fully one-half of the timber of the state." And then disaster arrived with the chestnut blight, a fungus that hitched a ride to the Bronx Zoo on nursery stock of Asian chestnuts, and then began to spread. In one of the most chilling collapses of a species, the blight eliminated mature chestnuts throughout its range from Georgia to Massachusetts by 1925. Rather than let the prized wood go to waste, a logging frenzy removed the dying trees in short order. The largest volume of timber ever harvested in Connecticut was 1909, its target chestnut.

With chestnut gone, and charcoal being replaced in forges and blast furnaces by coke made from coal, a greatly diminished Connecticut forest began to regrow in the 1910s. It had three times as much hardwood as softwood, with an age structure skewed heavily to young trees. By 1938, only 10 percent of the forest trees were older than sixty years. Half of the forest was between twenty and sixty years old, with only the older trees vulnerable to the wind. This is an entirely different forest from elsewhere in New England, and it set the stage for a different pattern of destruction. Even though Connecticut was hit first and thus hardest by Thirty-Eight's winds, its forest loss was mitigated by the fact that so much of it was too young to blow down. Since hardwood dominated the forest, it accounted for most of the blowdown. Twenty-year-old oaks and hickories withstood the gale and gave today's forest a substantial head start.

SIX

Cleaning Up

The dawn of September 22 ushered in two weeks of brilliant sunshine that illuminated for New Englanders their changed landscape. Street trees that had provided cool shade were tipped over, displaying their scraggly underpinnings. Familiar horizons showed startling gaps. In groves and backyards once tempered by shade, the light had become harsh. Leaves, twigs, branches, and limbs mixed incongruously in piles with shingles, roofing, and doors, the natural and the man-made all pried loose by the wind.

Trees crossed every driveway, lane, back road, town road, county road, and state highway. Today's broad, open, interstate highways were a distant dream, so even the heavily traveled roads were bordered by trees. The Depression had slowed down the economy, but it hadn't stopped it. The Northeast was still a center of manufacturing, and trucks had to get through to deliver raw materials and finished goods to and from factories in Connecticut, Massachusetts, and New Hampshire. People had to get to work.

That meant the first priority was to clear the roads. Throughout

New England, in cities, villages, and back country, people cleared the hurricane debris. "Today you could have done it quickly with a chainsaw, but we had to do it by hand with a crosscut saw," Dustin White said. "Everybody had a horse or a pair of horses because it was all little farms all over the area then, so they more or less worked on their own road. Once they sawed them into logs, they'd hitch onto them and the team would get them out of the road. They cut them up later because they were more or less in a hurry to get the roads open."

As White suggests, chainsaws would have come in handy, but they were still a decade away. Nor were there yet many tractors, the more useful ones being crawler tractors propelled on tracks like bulldozers and tanks. The Town of Corinth had one it used for plowing roads after winter snowstorms. "It took about a week to plow the whole town out with that crawler, and they ran it day and night. You could walk about as fast as it could go, you could walk right along beside it," White said.

In the rural areas, they piled the logs and brush separately alongside the roads, the brush to be burned and the logs to be sawn later into lumber or cut and split for firewood. As the roadside brushpiles multiplied and the sun sucked the moisture out of the wood, people started worrying about a follow-up catastrophe. While nobody alive at the start of 1938 had direct memory of a New England hurricane, fire was a different story. Many people had seen or smelled the smoke from forest fires that had charred the Northeast in the early years of the twentieth century.

The forests on Massachusetts's Cape Cod are particularly prone to fire. Indeed, pitch pine and oak dominate the forest in this low-lying sandy region largely because its dry soils and age-old history of periodic fire have removed the less fire-resistant species. Coastal areas had been the sites of a number of devastating fires since 1900, and as recently as April 1938 a wildfire had burned five thousand

In Barnet, Vermont, a man stands ready with axe and peavey before
tackling the first problem posed by the hurricane, clearing the roads so traffic
could resume. Courtesy of Dave Warden and University of Vermont's
Landscape Change Program

acres on the Cape and in Plymouth County and killed three firefighters. That was only five months before the hurricane hit, and
surely many in the Bay State had read about it in the papers.

New Hampshire, Maine, and the Adirondacks of New York hadn't
had recent flare-ups, but each had experienced difficult fire seasons
three decades earlier. In the summer of 1908, more than 300,000
acres burned in the Adirondacks, and 142,000 burned in Maine.
New Hampshire's worst year was 1903, when 504 fires burned
a total of 84,000 acres. Vermont, with fewer forested acres and a

greater portion of them in the relatively fireproof northern hard-woods, had its worst season in 1908, when 16,000 acres burned. These forest fires weren't on the scale of the conflagrations that routinely burn western forests each summer, but they were devastating on an eastern scale.

Smoke from these and earlier fires choked millions of people downwind, prompting widespread public demand for fire protection. By the late nineteenth century, timber companies had penetrated deep into the Adirondacks and northern New England's previously inaccessible hills and hollows, removing the large softwood logs and leaving behind a carpet of highly flammable needles, twigs, and branches. Timber barons, including J. E. Henry of Lincoln, New Hampshire, built railroad tracks into back country virgin timber where no rivers could be used to float the logs downstream. Loggers rolled the logs onto flatcars equipped with vertical bunks to hold the logs in place, and off they went to the mill. Thousands of miles of railroad tracks snaked through unpopulated areas, and sparks from steam-powered locomotives ignited railside slash. The steam boilers ran on wood or coal, both fuels capable of spraying sparks from the smokestack. In 1908 in particular, it was a cinch to map the line of fires that had started adjacent to railroad tracks. After decades of fires that could be blamed on locomotives, laws required that spark arrestors be installed on exhaust pipes. That had little effect because the spark arrestors didn't have a mesh fine enough to stop every spark. Besides, the $100 fine meant nothing to the large companies so the law was often ignored.

Government officials understood that people rather than nature were causing most of these eastern fires. Out west, fires start every summer from dry lightning, which happens when the air below the clouds is so dry that the falling rain from the thunderclouds evaporates before it reaches the ground. In the humid Northeast, rarely does lightning strike a tree without carrying with it buckets of rain

Timber companies built trestles and tracks to pull timber from the back country. Here is J. E. Henry's Engine No. 2 working on the East Branch and Lincoln Railroad circa 1900. The cars are loaded with logs from the Hancock Branch headed to the mill in Lincoln, New Hampshire. Courtesy of the Upper Pemigewasset Historical Society's Bill Gove collection

to extinguish any spark or smolder. Instead, people burning brush piles or singing around the campfire started many of the blazes, and in an era when most adults smoked, many fires resulted from someone tossing a cigarette butt or emptying a pipe of its dregs.

With fires at a crisis level in the Northeast, large-scale efforts at fire prevention and suppression began in the first decade of the twentieth century. Until then, states didn't have budgets for prevention, and firefighting was organized at the town level. The states created forestry divisions with a prime focus on fire control

throughout the region—Maine leading the way in 1891, Massachusetts and Vermont following in 1904, and New Hampshire in 1909.
Connecticut appointed a state forester in 1901, and fire control became a significant part of his job in 1905. Timberland owners in
Maine, New Hampshire, and Vermont created associations to help
protect their assets from going up in smoke.

They built fire towers, many of them as public-private cooperative efforts, knowing that a fire that was detected and contained
quickly was much easier and less expensive to fight. By 1938, some
320 remote mountains and hilltops in the Northeast sported fire
towers that were staffed during fire season and connected to civilization by telephone line strung through the woods.

Public education accompanied these advances in detection and
suppression, and extensive forest fires had become infrequent by
1938. Fewer fires were ignited, and most fires were contained before
they devoured large chunks of forest. Still, memories of smoke and
haze remained fresh, and the Cape Cod fire in April showed that
fire remained a threat. With all those new brushpiles, what was to
keep them from catching fire? And once on fire, what was to stop
them from spreading throughout the region? Thirty-Eight had destroyed much of the firefighting infrastructure. Many of the towers
had been knocked down, and the telephone lines connecting them
had been snapped and buried by falling trees. The trails to the towers were blocked, as were any roads providing access to the interior.
If a fire started in the back country, nobody could get to it quickly.
Not since the turn of the century had the people's ability to fight fire
been so compromised.

The fire protection system in place then and now was overseen
by state forestry divisions but administered at the local level. Appointed by the state on the town's recommendation, each town's fire
warden maintained the public safety by requiring permits to burn
brushpiles and working with the town's fire department, usually run

This farmhouse in Willimantic, Connecticut, sits exposed among acres of hurricane blowdown. It appears that the logs have been removed and the slash remains. Many people worried that ignited slash would start wildfires. United States Forest Service

by volunteers. Every fire warden knew his town's vulnerable areas, the shortcuts and which roads were most important, and where to find ponds and other water sources to fill their tankers.

Most of the fires three decades earlier had raged in the domain of the timber companies on remote mountains where few people lived. Now, in 1938, scattered piles of trees—most of them highly flammable white pine—were spread willy-nilly across the region, adjacent to houses and churches, many of whose steeples, incidentally, had been wrenched off by the winds.

Forestry officials differed on how great and how imminent a fire threat they faced. Austin Hawes was one of the many early graduates of the Yale Forestry School who filled important forestry positions at very tender ages, taking the position as state forester in Connecticut the year after his 1903 graduation. He served two tours in Connecticut sandwiched around a dozen years in the same position in Vermont. By 1938, he'd been the man in charge for nearly thirty-five years, a seasoned leader full of experience with forestry and with fire. Looking back at the hurricane response in his 1957 book, *History of Connecticut Forests*, he reported on a meeting held in Boston two weeks after the hurricane. Also in attendance was Ward Shepard, who had been appointed director of Harvard Forest following the sudden death of founder Richard Fisher in 1934. Hawes didn't bother with diplomacy in his assessment of Shepard's qualifications for the role: "There could hardly have been a worse appointment. Mr. Shepard knew nothing about forestry in New England and had had little training in silviculture. The hurricane destroyed the magnificent pine forest of the Harvard Forest, and director Shepard was on the verge of a nervous breakdown. Not realizing the difficulty of burning green timber, he conceived the idea that all of New England was in immediate danger of conflagration, and rushed to Boston to get the governor of Massachusetts to take immediate action."

Shepard was a Harvard Forest alum whose primary on-the-ground forestry work was in the high-desert forests of New Mexico and Arizona. More recently, he had worked in Washington, D.C., as a policy analyst and a public relations officer for the Forest Service and the Bureau of Indian Affairs. Those positions meant that Shepard had the ear of national officials, and he wanted their immediate assistance. Hawes reports, "The dramatic point in this conference was reached when John Foster, state forester of New Hampshire, stated in his calm way that he did not consider that there was any danger of immediate conflagration so long as the timber was green. Mr. Ward Shepard became so red in the face and so outraged I thought he would burst a blood vessel. He was indignant that a forester should make such a statement." Some foresters, proudly calling themselves dirt foresters, have mud on their boots and tree-marking paint on their vests and trousers. Others are marked only by ink and paper cuts. Shepard was the latter.

Hawes's depiction of Shepard's agitated state is fleshed out by a story in Harvard's student newspaper, the *Crimson*, in which the Harvard Forest director fed the writer some brimstone: "'The fires that are likely to result from the accumulation of these huge piles of tinder will be of a spectacular variety which has never been seen in this part of the country, usually being confined to the great timber stands of the West coast,' he said. 'They will be of the type known as "crown fires" where the flames shoot to a height of 500 feet, creating a mighty draught which throws burning brands half a mile ahead of the main blaze and makes it absolutely uncontrollable.'" The article concluded with a dire forecast for public safety: "Whether this area can be saved even with the strictest precautions observed seems doubtful to Shepard, and while planning for the safety of Petersham's 700 inhabitants, he more or less expects the town to be leveled to the ground along with an inestimable part of the surrounding country."

Shepard delivered that apocalyptic vision to the Boston meeting and to the opinion pages of the region's newspapers, foretelling firestorms racing through villages and swallowing everything and everybody in their path. He appealed directly to Forest Service officials, who had a vested interest in the White Mountain National Forest, where an estimated 200 million board feet had blown down on its 663,000 acres. A thousand miles of trails were blocked, and roads were crisscrossed by as many as two hundred trees per mile.

The damage was startling, but the national forest's loss represented only 5 percent of the estimated blowdown. Another bit of perspective: at that time, national forests comprised 175 million acres nationwide, making the White Mountain less than half of 1 percent of the total—not a size that would grab anyone's immediate attention. Unless, of course, it caught fire.

Unlike the vast stretches of national forests and Bureau of Land Management lands that dominate the western landscape, New England has never had much public land. Because the virgin forestland was settled and turned into farmland so quickly with the clear goal of having people out living on the land, the governments—first the British, then our federal and states—reserved almost no land for themselves.

It wasn't until the Weeks Act in 1912 allowed the federal government to buy back privately held land that government ownership of New England land increased. In the decades following Weeks, the federal and state governments began establishing publicly owned lands—state parks and forests, national forests—in New England. The White Mountain and Green Mountain National Forests were created in 1912 and 1932, respectively, much of the land being purchased from the delinquent tax rolls after its owners had cut its timber and stopped paying taxes on it once its timber value was gone. All of this land acquisition for the public made hardly a dent. Today across New England, individuals and corporations own more than 78 percent of the land. In Utah, only 30 percent of the land is owned by individuals or corporations; in Alaska, it's 11 percent.

What propelled the Forest Service to Boston was not its two national forests in the stricken area but the opportunity to influence the private owners who owned the vast majority of the down timber. The crusader Shepard wanted the government to control what private landowners did with their forestland. The fire danger was a convenient foot in the door for the government to improve industry practices. New England needed help, and Shepard wanted to be the savior. Fortunately, he had a good working relationship with the man at the helm.

The Forest Service chief was Ferdinand Augustus (Gus) Silcox, another of the early Yale foresters. In his first five years out of Yale, Silcox had so impressed his superiors in the Forest Service that he'd been promoted seven times. The seventh promotion brought him in 1910 to Missoula, Montana, as an associate regional ranger (second in command) of a region that encompassed fifty thousand square miles of the northern Rocky Mountains in Washington, Idaho, and Montana. Fires that year in that region changed the course of Forest Service history because it was the summer of the Big Burn, the biggest fire in the nation's history.

From his Missoula office, Silcox served as quartermaster, hiring and deploying firefighters and procuring and delivering supplies to those fighting thousands of small fires that had plagued the crews throughout that drought-ridden summer. The firefighters (including ten companies of army troops) had been able to keep the lingering fires from spreading until August 20, when a cold front blew through the Idaho-Washington borderlands with 75 mile per hour winds. For a full day, the huge winds turbocharged the smoldering fires, and they coalesced into a huge blowup. In her 1956 account, Betty Goodwin Spencer wrote, "Great red balls of fire rolled up the mountainsides. Crown fires, from one to ten miles wide, streaked with yellow and purple and scarlet, raced through treetops 150 feet from the ground." The inferno incinerated an area 120 miles long by 30 miles wide in forty-eight hours. The most deadly National Forest fire ever, it killed eighty-seven people, including seventy-eight firefighters. Those deaths seared Gus Silcox, and he believed with his entire being that if only they had the proper infrastructure in place—an extensive trail system, strategic fire lookouts, and plenty of manpower—those deaths and that level of destruction would not have happened.

Appointed twenty-three years later as FDR's forest chief, Silcox had the opportunity to right the wrongs of 1910, and he continued to believe that he could protect the nation's resources from fire if he had enough manpower. With four hundred Civilian Conservation Corps (CCC) camps built in national forests west of the Great Plains, he now commanded an army of foot soldiers (the CCC quickly became known as "Roosevelt's forest army"), and he put them to work building lookout towers, reducing fire hazards, and establishing trails and truck trails (a euphemism that circumvented the prohibition against building roads) into even the most remote country. When fires arose, the CCCers fought them alongside the Forest Service and personnel from the states.

Under Silcox, the Forest Service focused single-mindedly on beefing up its capacity to fight fire, and used its ready access to the CCC personnel to pounce on any fire that started. In 1935, Silcox instituted what has come to be known as the 10 A.M. policy. Every fire detected was to be extinguished by 10 A.M. the next day. His experience with the Big Burn had convinced him that any fire needed to be stopped immediately before it had an opportunity to grow into a real menace. And if you somehow didn't manage to control it today, your mission was to control it by 10 A.M. tomorrow. Immediate suppression of all fire was the goal, and it created and reflected a mindset that all fire is bad. Fire kills people and destroys resources that people need. The only good fire is a dead fire. The 10 A.M. policy was specifically rescinded in the 1970s, but its legacy of suppressing fire continues. And now that more and more people are living out west in what has come to be known as the woodland-urban interface, the impulse to suppress fires sometimes trumps the ecological understanding that fire is natural and it has shaped the forests of the high desert and intermountain West.

Ward Shepard used his Washington connections and got in the door to see Silcox with a delegation from Massachusetts on September 29. Shepard trumpeted the grave threat of fire, and Silcox, predictably sympathetic to their plight, suggested that Shepard and company develop an emergency plan. Silcox directed his regional foresters to cooperate with the state foresters to conduct an inventory of the regionwide blowdown. Meanwhile, the Massachusetts delegation developed its plan and presented it directly to FDR through Massachusetts Governor Charles Hurley on October 6. The Forest Service parachuted into action, creating the New England Forest Emergency (NEFE) program. To run it, Silcox appointed his head of state and private forestry, Earl W. (Ted) Tinker. As the chief's personal representative, Tinker left for the hurricane area on October 7, accompanied by several others from the Forest Service, setting up the emergency program's headquarters in Boston's Bellevue Hotel. When they started work the next day, Tinker received the foresters' statewide damage estimates, which pegged the blowdown at 3.7 billion board feet, a figure later reduced to 2.6 billion board feet. The foresters reported that the potential for fire had been enhanced by the sudden fuel load in 904 towns in 51 counties. One hundred of these towns faced extreme danger.

NEFE's appraisal of the situation in 1938 is summed up in an extensive report on the relief efforts written five years later by program administrator John Campbell:

> From a region of moderate fire hazard with comparatively favorable conditions for fire control, the area struck by the hurricane became a patchwork of blown down forests which, being mostly pine, represented a new high in forest fire hazard. The massed pine tops were highly flammable, and the tangle of fallen trees would have resisted any normal fire control efforts. This critical condition presented a problem of such magnitude that its diminution was considered more urgent than the problem of salvage. It not only threatened further loss of the log values involved in the blowdown but it constituted a major threat to remaining property and to human life. With several

weeks of dry weather followed by high winds, small fires might have been blown from one area of down timber to another and become widespread conflagrations, with imminent danger to everything in their paths.

How imminent was the danger? That continued to be debated. Foresters like Austin Hawes and John Foster counseled Tinker that the fire danger was not immediate and that it wouldn't be a big problem until the following spring. Fire season in the Northeast tends to be in spring, after the snow melts and before the trees leaf out. In that brief window, if the leaves and twigs in the understory dry out and someone gets careless with cigarettes or brushpiles, ground fires can get started. Long-term tracking of fires in the region shows April and May as the most fire-prone months. The threat of fire wouldn't last long because the most flammable fine fuels decay in a year or two.

The New England chapter of the leading professional association of foresters, the Society of American Foresters (SAF), distanced itself from the fear-mongering of Shepard and other alarmists and even suggested censorship to make sure that the public wasn't needlessly panicked. It wanted to prevent the publication of material like one widely read magazine article that said that a "fire once started could have roared through New England with greater velocity than the hurricane itself." The SAF felt that the aim should not be to frighten the tourist away from New England, "but once he is here we should scare the hell out of him so that he will be careful with fire in the woods."

Regardless of any forester's estimate of the relative immediacy of the danger, they all felt the need to regain their capacity to fight fire. Even Hawes and Foster concurred with that.

The citizenry's level of anxiety depended largely on how accustomed they were to logging. For years preceding the storm, New

England's lumber production had been concentrated principally in Maine and northern New Hampshire and Vermont. Where logging was routine, people were used to seeing large areas of slash left behind when the loggers moved on. People in the North Country hadn't forgotten that fires could and did occur, but they were more concerned with their loss of investment that lay on the ground. In Massachusetts, Connecticut, and Rhode Island, where the timber industry had withered, residents were unaccustomed to the jumble of logging slash, and every treetop and brushpile seemed to be just a spark away from ruining their lives.

That said, in the immediate aftermath of the storm, with the crisp, dry autumn making people nervous, the governors of New Hampshire and Vermont—Francis Murphy and George Aiken—took the unusual precautionary step of closing the woods to the public. While such a restriction is not unprecedented, governors rarely bar the populace from a walk in the woods. When hunting season opened in New Hampshire on October 14, the woods were silent because the threat of fire was so high. The ban was lifted on October 24 after rains had reduced the fire danger. Massachusetts also closed its woods in the central part of the state and was the only state to directly appropriate money to clean up the mess in the forest. After that, the weather was with them. A snowy winter provided lots of snow to melt in the spring, and March and April 1939 had plentiful rain, so the most vulnerable time passed by without any significant fires.

The New England Forest Emergency project established a primary goal of reducing the fire risk to prehurricane levels. The massed fuel posed a threat itself but also presented an impenetrable obstacle for reaching any fires. Minimizing the threat of catastrophic fire meant two goals had to be accomplished: rebuilding the region's capacity to fight fires and reducing the fuel load. In essence, the work was one and the same, disposing of flammable material. The first step was to reestablish the usefulness of the fire towers. That meant clearing trails to the towers and extricating the telephone lines from the tangle on the ground. Once that was accomplished, crews would open up all blocked roads, no matter how small. With roads open to vehicular traffic, they would then clear trails and fire lanes. All of these steps were to allow the towns' volunteer fire departments to reach any area they could have reached before.

The next priority was to remove flammable debris from the immediate vicinity of buildings and villages "where human life was most endangered." Next workers would focus on disposing of the brush they had piled roadside, removing any treetops within fifty feet of roads and trails. Doing so turned every road into a firebreak—a gap in the burnable material—that would starve any fire of fuel and stop it before it spread far. They never intended to conduct a clean sweep of the 600,000 acres of increased risk. Only those areas where the accumulated combustibles posed a great threat to human life or valuable property would be completely stripped of brush.

NEFE saw itself as a resource for the towns and states and avoided creating a top-down administration. Each state had its own director, and each state director organized his own work.

Local fire wardens scouted and mapped the worst areas. Local selectmen explained to the town's residents the danger and the need for cooperation. Given the relative scarcity of public land in New England, the bulk of the hurricane damage was on private land. Inattention to the hazard by an owner could threaten the property of neighbors doing their part to minimize the damage, but no government agency—federal, state, county, or town—could proceed with cleanup work without a signed release from the property owner.

Cleanup meant burning. No disposal site or collection of sites could handle the enormous volume of limbs and branches, so it was burned in place under controlled burning situations. In the areas where the more flammable white pine predominated—Massachusetts and New Hampshire—the amount of brush seemed staggering. Workers stripped the fine, more flammable material—the twigs, leaves, and branches—leaving the salvageable larger limbs and trunks looking like plucked chickens. The most effective disposal method was progressive, or "swamper," burning: the brush was cut and immediately thrown on a fire, eliminating the hazardous material in one operation. When rain or snow sufficiently reduced fire risk, swamper burning was the ticket. Otherwise, workers arranged it into tidy, dense piles for later. The structure NEFE recommended was known as a rick, an orderly stack of brush piled horizontally on parallel stringers, and contained by stakes driven into the ground. Crews arranged each layer of stems and twigs in an alternating direction so that the rick could hold a large volume. They capped it with dense branches to shed rain and snow, and when workers went back to torch it, the gap between the stringers fed air to the blaze so it roared. These ricks were designed for easy ignition and complete combustion, so no further cleanup would be necessary. Crews of twenty men could burn four hundred ricks in a day, and a competition quickly developed among the crews. The record was held by an eighteen-man crew in Vermont, which burned twenty-three hundred ricks in one day.

It was clear to NEFE's Ted Tinker that the town fire wardens, as good as they might be, couldn't handle the emergency on their own, especially in the towns with moderate or heavy blowdown. Franklin Roosevelt sent a memo to Chief Silcox on October 8 instructing him that NEFE should coordinate its efforts with the Works Progress Administration (WPA) and CCC. Silcox, whose tenure had always included extensive cooperation with these agencies, certainly didn't need FDR's urging, but the president's involvement lubricated the gears of interagency collaboration.

With these two relief agencies providing work for thousands of unemployed, a handy labor force went to work quickly. By November, there were 12,500 WPA laborers at work on hazard reduction, and by March 1939 the number had reached 16,000. The CCC was right behind them. There were forty-nine CCC camps already in place in the region, most of them based in state forests, populated by about 3,500 men in November and December and reaching a peak of 5,500 in March 1939. The CCC led the charge on reestablishing access to fire towers, clearing 458 miles of roads to them. The WPA, on the other hand, devoted much of its energy to clearing existing roads and fire lanes.

The WPA and CCC were both designed to provide work for as many people as possible instead of just giving families handouts, but they had entirely different cultures. To join the CCC in the late 1930s, young men had to be between the ages of seventeen and twenty-eight, physically fit, unemployed and unmarried, and willing to have twenty-five of the thirty dollars they earned each month sent back home. Everything about their daily lives was colored by the fact that CCC camps were run by the U.S. Army. The men lived in barracks and woke each morning to a bugle blowing reveille. From the start of construction, the army had command of the camps, including food and other supplies, administration, and medical care. The Forest Service or other supervising agency provided and planned the projects and supervised the work. The army's involvement suited the states' forestry divisions, which had considerably less manpower and budget and couldn't have handled the administrative responsibilities.

The CCC crews, made up of young men living in camps away from home, were already well trained for this sort of work. Campbell wrote that the CCC workers proved to be "excellently adapted to this project" because of their training and equipment and their availability. They did tremendous hazard reduction work, but "even this did not overshadow their inestimable value as a fire fighting

CCC boys wetting down hurricane slash in a hurricane-damaged spruce-fir stand during a demonstration of forest fire control. The site was along Cherry Mountain Road in the foothills of the Presidential Range in New Hampshire's White Mountains. United States Forest Service

force." These young bucks were hardworking, disciplined, competitive, and full of piss and vinegar, something that was probably never said of the WPA men.

Workers for the WPA were day laborers, and the program was administered by each county's relief office. If a family qualified for relief, the WPA would employ one family member with no consideration of that person's age. They weren't necessarily young, and they were generally unskilled. Pay was better than in the CCC, as it was intended to match local rates. Much of the WPA work was grunt work, manual labor improving local infrastructure, primarily roads and sewer lines. While the CCC had broad public support, the WPA operated in a more politically charged atmosphere. Critics saw many men standing around leaning on shovels. They saw vast bureaucratic inefficiency.

Worthy White, Dustin's father, was a selectman in Corinth at the time of the cleanup, and Dustin relishes retelling his father's story of working with the WPA. "The selectmen thought they'd get the WPA in because they had men available, maybe save the town some money and hurry up the job. So they got the WPA in, but just like anything that's got to do with the government, you got all this paperwork you've got to fill out. I can remember my dad sputtering about it. They had them help for a while and then a fella came along one day, one of the head ones, and he asked my dad, 'Where's your truck? Where's the truck on the job?' He said he had a bunch of fellas cutting out brush, and they didn't have a truck. Dad says, 'What do you want of a truck?' He said, 'If one of them gets hurt, we've got to take them to the doctors or somewhere.' Dad told him that the men in this town ain't so damn big that you couldn't draw 'em off in a car."

The WPA men were diverted from the road projects they were working on, and the Forest Service didn't have the personnel to supervise these laborers, many of whom lived in cities and hadn't spent any time working in the woods. You can get hurt working in a tangle of downed trees. Limbs under tension are just waiting for a release to spring back into place. Logs can roll, limbs can snap. It's unpredictable, challenging work. Campbell wrote, "In October 1938, the director issued for WPA a log cutting instructional circular designed to prevent damage to merchantable timber by the unwonted WPA axemen." It's hard not to also hear "unwanted" in his unlikely turn of phrase.

Contrast that assessment with the WPA's own portrayal of its hurricane relief efforts, as seen in an eleven-minute film produced in the hurricane's aftermath. *Shock Troops of Disaster* bears a striking resemblance to wartime newsreels, depicting feverish activity accompanied by charged music and stentorian narration. Referring to the WPA, the narrator described it in this way: "Manpower, turning from regular public improvements and services into the breach in times of dire need. 'Shock troops of disaster' someone has called them, because so many times in recent years they have provided the human sinews for great tasks of reconstruction." The film concludes with this gem: "The call may come from a great drought on the plains or a raging fire in the timberlands. Wherever there is trouble on a heroic scale, there you will find this peaceful army recruited from the ranks of the unemployed, working hard and well with the other agencies of mercy." Because it faced a good deal of public resentment, the WPA produced blatant propaganda like this through its own public relations arm.

In appraising the WPA's efforts, Campbell wrote: "Regarding the accomplishments of this working force, it should be stated that they contributed more man-days and accomplished more actual hazard reduction in terms of miles and acres than any Agency. They performed good work in accordance with the accepted policies and standards although their productivity was less than the CCC or [Forest Service] crews." Damning with faint praise, for certain, though ready access to that much manpower was an undeniable asset.

Another arm of the WPA, the Federal Writers' Project, provided another important hurricane-related accomplishment. Within one month of the wind and floods, it produced a truly fine book called *New England Hurricane: A Factual Pictorial Record.* Hundreds of photographs and eyewitness reporting fill the 220-page book, a remarkable documentary of the event largely free of any WPA self-aggrandizement.

NEFE ascertained that some of the worst hazards were in remote, sparsely populated areas beyond the range of the CCC and WPA crews. It developed commuting crews to work in these areas, paying the men fifty cents an hour. This proved very effective because it more closely resembled private employment—local men were hired to do the job, and if they didn't work out, they were let go. NEFE also began constructing work camps to be run by the Forest Service for fifty or a hundred men. Nine of these camps began operating in Vermont and ten in Massachusetts in June 1939, and their workers rivaled the CCC in terms of experience and training. Some of the men were forestry students; others were CCCers who had reached their maximum enrollment of two years.

Everywhere except on public land, these three federal agencies were working at the invitation of the town and in support of the town's fire department. By the time the work was deemed completed in November of 1940, they cleared 10,121 miles of trails and roads. They disposed of 214,902 acres of blowdown along roadsides, near homes and villages, and as firebreak strips on the edges of extensive blowdown. They reconstructed fifteen fire towers and reconnected 563 miles of telephone line. The total labor expended was just short of five million man-days, a remarkable investment of manpower that could have happened only under a New Deal. They contained any of the fires that started, none of which developed into "second-day fires," or what Campbell termed "serious and consequential fires."

The worst fire to follow the storm began not in the woods but at a sawmill processing lumber from the hurricane. In April 1941, with the woods dry and unseasonably warm, sparks ignited a saw-dust pile at a portable mill in Marlow, New Hampshire, and before it was contained, it burned twenty-four thousand acres in Marlow and adjacent Stoddard. Debris from the hurricane provided at least some of the fuel for the fire, which burned houses, barns, and other buildings.

At the close-out of the program, Congress authorized the transfer of the tools and equipment the Forest Service had purchased to the state forestry departments. A substantial amount of fire control equipment was thus given to the states according to their respective needs as a result of the hurricane, including twenty-five trucks, nine station wagons, twenty-two fire pumps, and 47,120 feet (almost nine miles) of fire hose.

Another of the NEFE administrators, Earl Peirce, wrote in his memoir, "The benefits resulting from the fire hazard reduction operations cannot be precisely evaluated. The protection taken certainly was instrumental in preventing any major fire losses. But how can the value of prevention be measured against the uncertainty of what might have occurred? An old proverb tells us that 'an ounce of prevention is worth a pound of cure.' Disastrous forest fires following in the wake of this hurricane would have raised this ratio many fold. Undoubtedly, the intangible but nonetheless greatest benefit was psychological—the lessening of local apprehension over threatening or possible danger."

SEVEN

Loggers and Sawyers

Congress was not in session at the time of the hurricane, so New England senators and representatives could see the damage first-hand. Population centers would have gotten their attention first, but in case they didn't understand how bad conditions were in the woods, their constituents let them know. One such letter found its way to the archives, written to Senator David Walsh of Massachusetts on October 5 by Judron Foster, a selectman in the Town of Westminster. Sparked to write by "numerous inquiries" he had received, Foster focused his letter squarely on the timberland.

> This disaster, to my mind, measured by value, is far greater than dwelling and factory damage combined, and it seems if the Federal Government is favorable toward performing a magnificent and far reaching service, ways and means of operating these damaged timberlots could be established. It is vitally important, however, that all the lumber be housed and protected until more favorable prices are assured.

The average rural inhabitant of New England has saved and nurtured his timberland to tide him over in later life, in other words, timberland has been his bank account. Many of these individuals haven't under normal conditions means to operate their lots, cutting, drawing, sawing, etc., say nothing of clearing out the present tangle of fallen trees, and it would seem that unless something definite transpires before spring, the fire hazard will be a second threat against the revenue derived from taxation.

In conclusion, I most earnestly ask you as the representative of the people in this vicinity, to make every effort to establish ways and means to cope with this most unfortunate condition.

People knew how to write letters back then (isn't "a magnificent and far reaching service" priceless?), and Foster sets the stage very well, noting the double whammy of the fire threat and further loss of income potential for a populace already reeling from the Great Depression. He and many others had heard stories of sawmill owners and loggers offering their services to forestland owners to remove their damaged wood at a price that was far below the going price paid for stumpage. Stumpage is the rather artless term for timber sold on the stump or "as is, where is." Before the hurricane, white pine stumpage was bringing between $4.18 and $6.79 per thousand board feet. An official from the Federal Land Bank of Springfield, Massachusetts, elaborated on the problem, noting that "the mill men, at least some of them, thought they were going to make a good thing out of the blow down. In some cases, they offered to remove the timber only if it was given to them, and in other cases offered from 25 cents to 50 cents" per thousand board feet. In southern New England, as Foster suggests, most of the forestland owners were not working the land themselves but were holding it as a bank account. So some unwary landowners were thus liquidating their bank accounts at a value of pennies on the dollar just to be free of the colossal mess in their woods.

The potential for gouging made Foster realize that the lumber would have to be stored long enough for its value to be stabilized, and he also realized that it couldn't be done without the "ways and means" of the federal government. Others who acknowledged the same situation suggested that logs rather than lumber should be stored while prices could stabilize.

The Forest Service analyzed the situation in this way. The average volume of lumber sawn in New England's mills in recent years had been five hundred million board feet per year. Reliable estimates showed there was at least five times this amount lying on the ground. The lumber industry, having contracted during the Depression, simply couldn't build itself back up to handle this amount of timber in a timely fashion. And even if mills could saw it, who would buy it?

The market couldn't be counted on to take care of the problem, because any increases in normal market activity would handle only a small amount of the downed timber. Further, as long as supply exceeded demand so vastly, there would be little or no price paid to the landowners for stumpage, and negligible profits for loggers or manufacturers. In addition, the Forest Service officials were concerned about the consequences of failing to salvage the wood. They feared that the "remaining down timber, for which there was no immediate market, would rot on the ground, forming tremendous insect, disease, and fire hazards in addition to economic loss."

The Forest Service saw the need for a stabilizing influence on the price of logs and the flow of lumber to market. The only way it could accomplish that was to put the power of the federal government to work, establish a fair price for logs, and buy up all it could. The service could then meter the logs out into the market as demand required. At the heart of this reasoning was that the purchasing program would allow thousands of local landowners to realize a decent return from what could have been a nearly total economic loss.

The only problem was that the Forest Service had no authority to purchase two billion board feet of logs, or the money to spend on it.

■

From our vantage point today, the U.S. Forest Service seems like a venerable institution with a long, rich history, but let's not forget that it did not exist until 1905. That means that when the hurricane hit, it had been operating for a little more than three decades. The agency was the brainchild of Gifford Pinchot and Theodore Roosevelt, and it was conceived with a lofty mission: "Where conflicting interests must be reconciled, the question shall always be answered from the standpoint of the greatest good of the greatest number in the long run." America's public land serves many different purposes, so the conflicts among the users—ranchers, loggers, miners, hikers and others in search of recreation and beauty—have long been complex and contentious. Pinchot's formulation of the agency's work started with philosopher Jeremy Bentham's "greatest good for the greatest number" but added the final phrase "in the long run" because forests have life spans that dwarf that of humans, and foresters are trained to think well beyond the present stand of trees. The Forest Service is most widely known as the administrator of the nation's many national forests, but from the start it has had two other core responsibilities: to conduct research and to assist and cooperate with administrators of state and private forests.

Teddy Roosevelt had Gifford Pinchot as his chief of the Forest Service, the two men sharing a conservationist spirit. The second President Roosevelt is less well known as a conservationist and outdoorsman, partly because by the time he entered the national spotlight, he'd been confined to a wheelchair by polio. But like his older cousin, FDR loved the woods, and he too had been influenced by Gifford Pinchot. He personally planted thousands of trees on his own estate in Hyde Park, New York, and he saw forests as providing solutions to a number of problems. As president, he proposed to battle the loss of topsoil in the Dust Bowl by planting a shelterbelt of trees extending from the Dakotas to Oklahoma. And his "forest army," the CCC, was planting trees, battling against insects and diseases that plagued trees, and, of course, suppressing fires. For his chief, Franklin Roosevelt chose Gus Silcox, a perfect match with the New Deal, both philosophically and administratively.

After the horrors of the Big Burn, Silcox set about building the best Forest Service region ever, and by all accounts, he succeeded. He was both a systems guy and a man's man. People trusted him. In Missoula, he mediated a dispute between the lumber companies and the lumberjacks, 90 percent of them Wobblies, members of the Industrial Workers of the World. Well educated but not from patrician stock, Silcox had deep respect for the working man. When the United States entered World War I, his administrative and diplomatic skills were put to use to in the Seattle shipyards, where shipbuilding was progressing at a crawl because of disputes between labor and management. Summing up his time in Seattle, he said he "prevented labor from lying down on the job and management from perpetuating grossly unfair hiring and firing methods." He had compassion for the have-nots struggling through the Depression, and he was adept at finding administrative solutions to the problems that faced his agency and the nation.

Silcox was a visionary and a crusader who believed that the forests and their resources were held in trust for all of the people. In his 1937 letter to his Forest Service people celebrating Christmas and New Year, he wrote, "We are still up against the paradox of poverty in the midst of plenty. Except for industrial centers, our most

Ferdinand Augustus "Gus" Silcox, forest chief of the United States Forest Service, stands in his office but looks ready to go put out some fires. Silcox was the author of the Forest Service's "10 A.M. policy," which dictated that any reported fire must be contained by 10 A.M. the next day. The controversial policy was officially rescinded in the 1970s, but the culture of stopping every fire has not disappeared.
Forest History Society

serious social problems are now in forest regions from which future citizens will in large measure come. We must rededicate ourselves to a broader public service. As trustees we must manage the Nation's forests so they may become tools—and better tools—in the service of mankind."

Hand in hand with his interest in serving mankind was his disgust with the forest industry, which he saw as pillagers of the public good for their own private profit. "After 30 years of [Forest Service] preaching, forest devastation has not stopped," Silcox wrote. He struggled to solve that problem, and his solution had three facets: acquisition of forests, cooperation with private landowners, and public regulation of forest practices on private land. To that end, he had been instrumental in drafting legislation that would have granted his agency regulative authority over timber harvests on private land. The measure had been defeated, but his efforts had made him many enemies in the forest products industry.

In the 1930s, there was a name for people who thought as Silcox did: communist. And indeed, after five years on the job, in the month before the hurricane blew through New England, Silcox was charged as such during a congressional hearing by Representative Noah Mason, a Republican of Illinois. Targeted as a communist, Silcox found himself in the company of seven other government officials, including Bob Marshall, one of the true giants of the American conservation movement who then worked for the Bureau of Indian Affairs. Silcox's reputation as anti-industry was well established at the time that New England's landowners faced the daunting dilemma of salvaging value from their downed timber.

So when Silcox designated a deputy to manage the New England Forest Emergency project, he turned to a man who was on better terms with the forest industry. Ted Tinker had been supervisor of two national forests, then worked on land acquisition in the Lake States of Michigan, Wisconsin, and Minnesota. Later, when the Lake States became a Forest Service Region unto itself, he was named to lead it. Success in that role brought him in 1935 to Washington, where he was named the first head of the State and Private Forestry Division. By all accounts, Tinker was a pragmatist, someone who knew how to get things done. Years later, a colleague, Leon Kneipp, recalled working with Tinker.

"He was a small man. He couldn't have weighed over 140 or 150 pounds at the most. On the Superior forests, you traveled by canoe. There were no roads in those days. Usually three men would travel in a canoe. They usually carried four packsacks; so when we came to the fords, one man would take the canoe, and each of the other two men would take two of the packsacks and carry them across. Sometimes I felt that the game wasn't worth the candle, but I'd see Tinker trotting along carrying his eighteen-foot canoe, and I would think, 'Well, if that little son-of-a-gun can do it, then so can I.' So he built my morale up more than almost anybody else could."

And Kneipp didn't fault Tinker for what might seem like callousness. "But just typical of Tinker, when we started out in the canoes, he made this casual remark. He said, 'You want to understand now that if anything happens to these canoes and they tip over, then it is every man for himself.' He said, 'A man won't live more than a half hour or an hour in this ice cold water, so he had better spend whatever time he's got in saving his own life instead of worrying about trying to save somebody else's.' "

In a telling summary of Tinker's work style, Kneipp said, "He had a very fine, keen mind. He was not overly sentimental or overly scrupulous or anything of that kind. I do not mean that he was obstreperous, but if there wasn't any law against it, then he didn't see why he should not do it."

Tinker worked well with the state foresters and with the forest products industry. He had run interference for his more idealistic boss before, assuring the industry that any regulation would be local, not federal. And even though he could have been deemed suspect because he'd spent much of his career in land acquisition for the Forest Service, Tinker had plenty of experience working with private landowners in the Lake States district, where the private acreage greatly outweighed that of the public, and he'd earned their trust. In a 1954 profile in *American Forests*, Nard Jones wrote about Tinker's assignment to lead the New England emergency effort, "It was a big job, a tough job that had to be done in a hurry. And it was fraught with political dynamite. Ted Tinker got that job."

At the same time he was overseeing the buildup of the New England firefighting capacity, Tinker was making the rounds, meeting with all the state foresters and with representatives of the industry. He sought their advice and their help in analyzing the extent of the salvage problem and what should be done about it. That said, there was never any doubt that the Forest Service would step in to buy much of what had blown down.

Behind the scenes, he and his boss were figuring out how to accomplish all of this without an appropriation from Congress, which was not expected to return to Washington any time soon. Neither the Forest Service nor the Department of Agriculture had borrowing power, so they had to come up with a different source of funds.

What Tinker and Silcox emerged with was a plan for the government to set up and fund a new agency to purchase logs delivered to storage sites it would set up. By doing so, it would control the vast glut of logs provided by the storm and keep them off the market until they were needed. This would stabilize the existing markets and would ensure that the glut created no serious disruption of the existing lumber market. Over time, the agency would sell the logs to industry at a fair price that would allow it to repay the loan that established the agency. It would cooperate with the industry rather than compete with it, and help it to pull itself out of the doldrums of the previous decade. And because the Forest Service always had an evangelical side to it, it would introduce to landowners and the industry a new approach to forestry that protected and improved the forest resource.

A memo from the New England Forest Emergency office on October 9 contained a report to the chief of the Forest Service. It estimated that 4 billion feet of timber (rounding up from the foresters' early estimate of 3.7 billion) was on the ground and that 60 to

Leon Kneipp, left, and Earl W. "Ted" Tinker. Kneipp said of Tinker, the
pragmatist who led the Forest Service's hurricane cleanup efforts, "He was not
overly sentimental or overly scrupulous. . . . If there wasn't any law against it, then
he didn't see why he should not do it." Forest History Society

70 percent of it could be salvaged. Since most of it was white pine,
it had to be manufactured into lumber or stored in water before
summer to prevent serious blue stain or insect damage. Blue stain
is a fungus that spreads into logs if they sit too long in the warm
weather, discoloring the lumber. The white pine sawyer beetle is
the primary insect that feeds on downed white pine, riddling it with
tunnels that make the wood unsalable.

The memo writer (it's unsigned, though it's surely from Tinker)
wrote that "neither the state governments, established timber industry, nor individual landowners are in a position to handle this

amount of timber within the time limits involved." He noted that an offer from the Farm Service Administration to make loans to individual farmers, while welcome, would not solve the problem.

He proposed to make use of the Reconstruction Finance Corporation (RFC), an entity wholly owned by the federal government that administered aid to state and local governments and made loans to private entities. He suggested that the RFC could make available a revolving loan fund of $10 million to a newly formed entity to purchase delivered logs from landowners. The new program wouldn't buy stumpage—it would be up to the landowner to arrange and pay for the cutting and delivery of logs. The stored logs would serve as collateral for the loan. The program would then sell the logs, or if it proved necessary, saw logs into lumber and sell the lumber.

What started out as a report quickly turned into a proposal and a sales pitch. The writer assured the chief that it would require a minimum of administrative expense. His team, the one being created to handle the problem, could take care of the details. And he offered to use the Boston office to get the program started and then hand it over to the loan maker over time.

What the government was doing, he argued, was to "merely establish a market for logs, and leave to the local people the business of logging and milling." Many owners of small woodlots would have preferred not to have to contract with the logger to cut and deliver the logs, but Tinker felt that it would be too cumbersome to get in the logging and sawmilling business. Four days later, on October 13, Silcox wrote a memo to Jesse Jones, the powerful chairman of the Reconstruction Finance Corporation. Following Tinker's recommendations, he proposed establishing a government-owned nonprofit corporation to handle the situation in New England. Management would come from the Forest Service, with broad powers to set prices, purchase logs, and sell logs and lumber. He pushed for a speedy resolution: "Farmers in this area are now being offered absurdly low prices for stumpage and it is of great importance that announcement be made at the earliest possible moment to prevent further exploitation of these people and a runaway market with a consequent demoralization which may spread a good deal further than the New England area."

The proposal was batted around, tweaked, and ultimately approved in short order, a display of inventiveness and flexibility characteristic of the New Deal environment. Silcox and Tinker created and funded a new program to handle the salvage by finessing relationships among three corporate entities that were wholly owned by the federal government: the Federal Surplus Commodities Corporation (FSCC), the Disaster Loan Corporation, and the Reconstruction Finance Corporation. Out of the fancy footwork came the Northeastern Timber Salvage Administration (NETSA), staffed with Forest Service personnel but operating within FSCC. The $15 million loan that established NETSA was secured by mortgages on all the timber it would purchase from landowners. The loan would be repaid out of proceeds from subsequent resale of the logs and wouldn't come due until January 1, 1942. Silcox quickly delegated his authority to Tinker to run NETSA. The pragmatist and the idealist clearly worked well together. By November 14, 1938, everything was in place.

The first item on Tinker's to-do list was determining the price to be paid for logs. NETSA studied price reports from the preceding five years and came up with a schedule of prices for the different species. The initial proposal had landowners receiving 80 percent of the price up front, with a supplemental payment coming after the government had recouped its money. Landowners found this unacceptable, and few of them delivered logs on this basis. NETSA responded by increasing the up-front percentage to 90 percent and increasing the prices paid for the lower grades, a concession large enough to start the trucks rolling.

A more difficult sticking point was deciding how to measure the logs. Given that a log is essentially a cylinder with a slight taper, it seems that knowledgeable men could come up with a reasonable method for measuring its volume. Over the years, they have—dozens of them. Each state had its preferred log rule or measuring system. In New Hampshire, they used the Blodgett rule, and across the river it was the Vermont or Humphrey rule. Maine had the largest selection to choose from, the Holland, the Bangor, the Saco River, and the Square of Three-Fourths. Massachusetts had its Mill Tally Log Rule for White Pine, and the foresters working on the Green Mountain National Forest used the Scribner Decimal C. The scaling systems are called log rules because the volumes and the corresponding diameters and lengths are printed on a long wooden ruler, more like a yardstick, used in the log yard.

The peculiarity about log rules, and this continues today, is that the log is scaled not according to the volume of a tapered cylinder but by how many board feet of lumber can be sawn from it. The rule predicts the volume of the boards that will be sawn from the cylinder. Each rule springs from a different formula that takes into consideration the unusable parts of the log: slabs cut from the perimeter and the sawdust lost from the saw's kerf or thickness. Some rules favored larger logs with little taper; others, the Blodgett, for instance, have a built-in advantage for logs with smaller diameters. Each state wanted to use its own rule because that's how its mills and loggers were accustomed to buying and selling timber. As the primary buyer, NETSA wanted to simplify its program by having a single log rule govern the whole region. After considerable study, it adopted the International Quarter-inch Rule as the standard, having determined that it would closely and fairly approximate the amount of square-edge lumber that could be produced by an efficiently operated mill. A tally of logs on the landing could be expected to consistently produce a corresponding volume of lumber; there would be no overruns that favored the buyer or underruns that favored the seller.

That solution satisfied nobody, and it absolutely enraged the boxboard industry, because it sawed round-edge, not square-edge, lumber. In sawing round-edge (which today is usually known as live-edge), they sawed each log straight through, never turning it on the carriage to find the best face. Straight-through sawing essentially does to a log what an egg slicer does to a hard-boiled egg. The boards' edges were left curved and intact, the bark often still attached. Because there was so little waste, the log volumes in the Massachusetts mill tally ran 10–20 percent higher than ordinary rules. Especially in Massachusetts, but also where boxboard still ruled in New Hampshire, the NETSA log rule became known as the "swindle stick."

The rest of the country had moved on to square-edge sawing and had developed a complex system for grading both logs and lumber. As Richard Fisher had demonstrated a decade earlier, New England sawyers needed to pay more attention to the grade of lumber coming off the mill, and New England forestland owners needed to pay more attention to the grade of log going into it.

NETSA pushed the New England industry along by imposing a grading system on the logs it purchased. This was tough love for log sellers, providing higher prices for the best logs, though there were few. In hiring log graders, NETSA favored young men out of forestry school to do the scaling because they were new to it and didn't have any preconceptions. NETSA paid eighteen dollars per thousand board feet for the highest grade pine log, fourteen dollars for the middle grade, and only twelve dollars for the common log. Seventy percent of the logs were purchased at the lowest grade. On top of that, logs that weren't straight had their scale reduced.

Jim Colby still fumes about the abrupt change in the way of doing business: "Good god, they were screwing the people right and left.

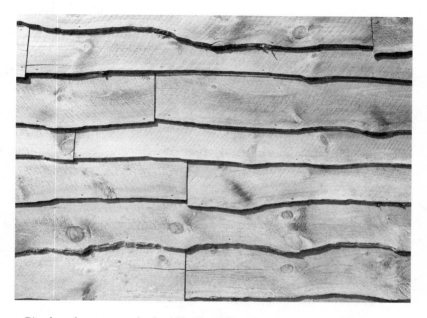

Pine boards sawn round-edged like these were the rule in New England in the 1930s. Not many sawmills saw round-edged lumber today, although sometimes you'll find it used as it is here, as siding on a rustic building. Today, it is often called live-edged. Sarah Smith

You'd have a log with just a little sweep in it and they would cut it right down as far as scale."

NETSA had some other restrictions. It agreed to purchase only from storm-damaged areas, and only those trees damaged in the hurricane. To ensure that it wasn't buying only the dregs from a woodlot, NETSA insisted that a landowner sell it all or none of his logs. The Maine state project director wrote, "We are not in the market for the 'skim milk' in instances where some private purchase has taken the 'cream.'"

Where did all these logs go? War movies often depict situation rooms in which generals examine a map and move pins around on it to show strategic positions: enemies, fuel depots, key topo-graphical features. The equivalent of that situation map for Thirty-Eight resides in the archives at Harvard Forest. A flat wooden box about three feet square, it unfolds on hinges to show a map of New England. Peppering it are hundreds of pins, their tops color-coded to designate either a NETSA or a private sawmill. Many of the NETSA pins also bear a flag with an identifying number. Dots colored red or blue mark the location of the log storage facilities, with blue signifying a wet site (pond or lake) and red a dry site. The density of the pins gives a clear impression of the extent of the damage. Central Massachusetts east of the Connecticut River looks like a dog's nose full of porcupine quills, southern New Hampshire slightly less so. Other areas are more sparsely marked. Few dots or pins adorn the map west of the Green Mountains in Vermont, and while they don't extend far into Maine, there is quite a clump of them in Cumberland County.

There are three pins in Fred Hunt's Rindge, three in Peter-sham, and a handful near Boscawen. In Vermont, Bradford has two, Corinth and Chelsea one each. The presence of a pin and/or a dot means that there was enough downed timber within ten miles or so that it was worth setting up a place to store and saw logs.

NETSA scouted for ponds because wet storage would buy them time, keeping the pine logs from staining or being chewed by bee-tles. Hardwoods don't float, so those would have to be stored and sawn quickly at a dry site, but pine and spruce could be stored al-most indefinitely. An early memo from Tinker shows that he as-sumed they'd be storing pine logs for as long as five to ten years. NETSA ultimately found 260 wet sites and began taking deliveries of logs, the first loads arriving at two sites in New Hampshire on November 21, 1938, one week after Tinker took charge of NETSA and exactly two months after the hurricane. A mythology has de-veloped over huge pine logs still to be found in the depths of New

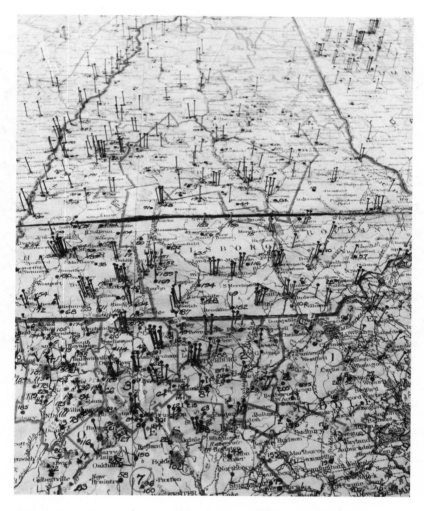

Each pin on this map represents a storage site and/or sawmill for logs salvaged from the hurricane. Think of it as a depiction of the relative brunt of the storm. The thicker the pins, the more blowdown that was salvaged. Stephen Long

Logs delivered to Connor Pond in Petersham, Massachusetts, from the blowdown at Harvard Forest. Two and a half million board feet were sawn by the mill set up at the pond. A. J. Lute, 1939, Harvard Forest Archive, Petersham, Massachusetts

England ponds. It's true that some of the logs sank or were stuck in the mud bottoms. NETSA even did a study of it and found that approximately 2.5 percent of the logs sank if they were stored for two and a half years. I spoke with a man who earned money for college scuba diving for logs from a New Hampshire pond, and he said that the logs were well preserved but tended to be small.

It didn't take long to realize that NETSA wouldn't be able to find enough wet sites, so it contracted with farmers to lease open land for dry sites, and at the dry sites there was no time to waste. The logs would have to be sawn by the middle of the following summer to retain their value. Tinker hadn't wanted to get into the business of sawing lumber, but the clock was ticking and NETSA began contracting with sawyers.

Many New England lumbermen felt threatened by the government competing with them, and rumors spread about NETSA planning to dump low-priced lumber into a market that had finally been stimulated by the need to repair and rebuild after Thirty-Eight's wholesale destruction. Tinker and his Forest Service crew met with industry leaders, including the heads of two trade associations—the Northeast Lumber Manufacturers Association and New England Lumbermen's Association—at the Toy Town Tavern in Winchendon, Massachusetts, and the working group became known as the Winchendon Committee. It hammered out an agreement spelling out the terms of the sawing program and publicized it to quell the rumors. Logs would stay in the ponds indefinitely, and only the logs at the four hundred–plus dry sites would be sawn. After sawing, the lumber would be stored at the mill site until a buyer could be found for the entire site. NETSA mills sawed the lumber square edge because it was to be marketed outside of New England—south to the Potomac and west to the Mississippi—so it wouldn't compete directly with the New England mills, which would presumably be purchasing the logs from the ponds to saw into lumber for local markets.

NETSA advertised widely for loggers and owners of portable mills to come to New England, and it was soon under contract with sawyers who trucked in portable mills from throughout the region and beyond, including New York, Pennsylvania, Michigan, Wisconsin, and West Virginia.

Joe Colby retrieved his mill, which had sat idle for the previous decade in Northfield, New Hampshire, though not without some drama. Colby and his crew spent a day freeing it from the trees that had grown up in and around it, and he saw that it was missing a few of its important parts. He knew that there was another mill set up about a mile from there, and he quickly put two and two together. His son Jim told me about going with his father to discuss the matter with the other sawyer. "This man from Antrim was sawing and he had a bunch of guys that stayed there in the house. It was a Saturday night, and they'd all been putting the booze to it, and we went in the house and my dad said to him, 'Joe, I'm gonna set my mill up in the next month or so and I'd like to have you bring back my sawyer's favorite, and the Moody compound, and the governor for my engine.'

"My dad was a good sized man and so was this other guy. He walked right up to my father, I can almost see his nose touching my dad's. 'You goddamned old potbellied bastard,' he says, 'I didn't steal anything from you.' My dad, he was pretty careful about what he said. He said 'Joe, I didn't say you stole it. I don't need it till a month or so. When you get around to it, you bring it back.'

"The guy screamed back, 'You get the hell out of here, and don't you ever come back here again.'

"I was scared as hell there was going to be a fistfight. We got out in the car, and my dad says, 'He'll bring them back. I know darn well he took them.'

"A couple of weeks went by and he never brought them back. So we went up again. I didn't want to but I went in with him. He says, 'Joe, you haven't brought those back yet.'

"'No, goddamn it, I told you not to come up here. I haven't got them.'

"My father had this metal stamp, it said J. G. Colby, so he said to Joe. 'I've got every one of those parts stamped. You wouldn't want the sheriff to come out and go through your mill and see those stamps, would you?' And this guy's face dropped about a foot. We got them all back, no sheriffs or anything."

The Colbys set up that mill in Concord's Rollins Park, whose old pines had been destroyed. Joe Colby knew a good thing when he saw it, and he quickly bought another mill, then another, until he ultimately had seven mills sawing up hurricane logs. He set up mills at dry sites in Canterbury, Boscawen, and Concord. He also sawed the logs stored in Clough Pond in Canterbury, Deermeadow Pond in Epsom, the Bay in Salisbury, and at the Girl Scout camp on Lake Tarleton in Piermont.

At two of the mills, Colby sawed square-edge lumber for NETSA at its going rate of $7.50 per thousand board feet. But at the other sites, he sawed round-edge lumber. At Clough Pond and Deermeadow Pond, he sawed on contract with New England Box Company, which bought the logs at the NETSA scale and made out well. Jim Colby explained how that worked. "If there was 100,000 board feet in the pond scaled by the government, it overran 25 percent, and New England Box got that 25 percent because the government sold them whatever was scaled in that pond. It worked out for us because we got paid so much per thousand to do it, and it worked out for them because they got the extra money. We did the same thing with the City of Concord. My dad was a very foresighted man and he went before the council and told them, saw this lumber, stick it up the way it should be stuck, and you'll make some money. That's exactly what they did." So it turns out that the swindle stick worked to the box companies' benefit.

Jim Colby served as "the run around guy. I would go from one mill to the other, make sure it had enough oil, or pitch in if there was a man shy." All told, the Colbys' mills sawed more than fifteen million board feet.

Over in Vermont, which had fewer ponds to choose from, one of the dry sites was in Bradford at Wentworth Blodgett's, who leased eight acres to NETSA. It had to be that large to store logs, set up a

Sawing square-edge lumber at the McGinnis sawmill from timber downed at Harvard Forest. Willett Rowlands, 1939, Harvard Forest Archive, Petersham, Massachusetts

sawmill, store sawn lumber, and store sawdust and slabs. The rent was a dollar a year plus a portion of the sawdust and slabs. Logs were skidded there from the Blodgetts' and three other neighboring farms. I could find no records to indicate whether logs were delivered by truck to the site from other owners in the vicinity, but given that Bradford and Newbury had significant blowdown, we have to assume that it bought logs from beyond the four farms.

Fifty years later, Put Blodgett's curiosity about the huge event from his childhood led him to Ken Stockman, who had helped salvage the logs. Stockman responded to Blodgett with a letter explaining how it went. His father, Harold, ran the operation for the Blodgetts and the three other farmers. He hired five or six men to go into the woods with their axes and crosscut saws. The workday lasted nine hours, which earned a man three dollars a day. Blodgett's father purchased an Allis Chalmers Model M crawler tractor designed for hauling logs, and Ken Stockman was given the job of running it. The crew started in late October, and the cutters worked their way through the tangles of trees and began to convert what they could into logs. Another man, a Finn from Corinth, used two horses to skid the logs out to skidways where they could be loaded on scoots for the crawler to pull to the landing.

A scoot was essentially a sled with two eight-foot-long elm runners and a pair of heavy bunks that could hold eight hundred to a thousand board feet of pine. Stockman would deliver an empty scoot to one of the skidways and then go pick up a loaded scoot at another. As driver, he didn't have to load the scoot, but he did have to unload it at the landing. He wrote, "In order to keep up with the cutters, there was no time to loiter. On the trip back from the landing you opened the throttle wide open and let it go." Stockman guessed that in six months, they salvaged between half a million and three quarters of a million board feet, "not bad for a bunch of mostly amateurs and considering the conditions we had to work under." It turns out that he underestimated; NETSA records show that it purchased 849,000 board feet at Vermont's Dry Site number 10.

Stockman's crew finished logging in April as mud season kicked in, just after a man arrived from New York with a portable mill. This was typical of the way the work progressed everywhere. Across the region, any man with the slightest logging experience could find a job. Besides the dangerous work, they suffered through a snowy winter. Jim Colby worked with the crew at Lake Tarleton Girls' Camp that winter. "We had thirty-four men chopping and logging, and four teams of horses. We'd shovel our way in in the morning

A crew headed by William Butler of Wrentham, Massachusetts, poses after delivering a fully loaded scoot pulled by a crawler tractor. E. A. Hanson, 1939, Harvard Forest Archive, Petersham, Massachusetts

and get in there about eleven o'clock and we'd shovel our way out at night. Oh, that was a terrible winter," he said.

Despite the difficulties, most of the logs were delivered by July 1939, with a peak of 110 million board feet delivered in March 1939. NETSA continued to accept logs until June 1940, at which point it had purchased 690 million board feet from 13,563 landowners, paying out $8.3 million. New Hampshire, whose citizenry has long prided itself on its independence and its belief in small government, was by far the largest beneficiary of the program, its landowners receiving 59 percent of the $8.3 million. The volume salvaged privately couldn't be measured precisely, but NETSA estimates that it came to 560 million board feet. Combining NETSA and private operators, 1.25 billion board feet was salvaged, slightly less than half the estimated 2.6 billion that had blown down.

Though NETSA hadn't wanted to be responsible for sawing lumber, in the end it managed to sell only 84 million feet—12 percent of the total—as logs. Along the way, it found itself with a huge standing inventory of square-edge lumber and nobody to buy it. Realizing at the start that it might own the lumber for quite some time, NETSA required the sawyers to sticker and store the lumber in piles, with overhanging roofs built according to its exacting specifications.

Across New England, in as many as four hundred farm fields and log yards, lumber stacked twelve feet high in pile after pile stood drying out, while NETSA tried to sell the contents of each yard in its entirety. Ultimately, it did even better than that, selling 425 million board feet—more than two-thirds of the lumber—to a single entity. L. Grossman and Sons, a pioneer in the building supply retail business in Quincy, Massachusetts, created the Eastern Pine Sales Corporation to purchase lumber and eventually logs. It signed the contract in late 1940, which was a huge relief for the NETSA officials, who could now go about their business knowing that they could make good on their loan. The total cost of the salvage program was $16,269,300, of which almost $15,000,000, or 92 percent, was recovered by the government.

The efforts of individual landowners were bolstered by an army of recruited loggers who brought the logs out of the woods. Hundreds of sawmills and their crews sawed it into lumber. CCC, WPA, and Forest Service crews stripped the logs of any highly flammable material. At a time when jobs were scarce, rural New England was put back to work.

Neither Silcox nor Tinker, the two architects of the operation, was around to see it all play out. Tinker resigned from the Forest Service on December 16, 1939, to take a job as executive director of the American Pulp and Paper Association. Only four days later, Silcox died of a heart attack. He had already written his annual holiday letter to his Forest Service family, this one under the title "Guarding Democracy." In it, he wrote,

Although these are days when armies march as dictators command, America stands firm for democracy. It is the job of every one of us to help maintain that stand. As a Nation we draw civic and spiritual guidance from the Declaration of Independence and the Constitution. For most material things on which our strength is based we turn to the earth, its minerals, its soils and waters, and to the plant and animal life they yield. As members of the Forest Service we therefore rededicate our efforts to securing wise use of our natural resources. For sources of raw materials, of necessities of life, and of employment and income, using natural resources wisely and well is fundamental to national defense against military aggression and against the undermining of economic and social structures within our borders. But abuse and depletion of natural resources are not the only threats to democracy as we know it. Freedom must also be guarded; freedom to seek the truth, and courage to apply it without prejudice or rancor through established institutions in defense of human rights. You and I are members of an organization permeated by the spirit of public service. Foresters, we are also citizens of a democracy. I am confident, therefore, that our efforts and our lives are also rededicated to preservation of tolerance, kindness, and those ideals that guided our forebears when, seeking blessed sanctuary, they founded this United States of America.

The day after Thirty-Eight transformed the region, the head-
lines in the papers told of its devastation and of Hitler's annexation
of the Sudetenland in Czechoslovakia. Then as the New England
Forest Emergency program reduced fire hazards and the Northeastern Timber Salvage Administration bought logs and sold lumber, the drums of war were beating ever more insistently. When
the United States entered the war in December 1941 following the
attack on Pearl Harbor, a huge percentage of the hurricane lumber was sold back to the government for military use, much of it as
crating. The government had gone out of its way to avoid the New
England tradition of sawing round-edge lumber for boxboard, and
then ended up using much of the nicely manufactured square-edge
lumber to make boxes and crates for shipping war materiel.

If John Campbell saw that irony, he glossed over it in touting
NETSA's accomplishments: "That New England, the nation, and
the anti-Axis allies benefited from the program is unquestionable.
The stock pile of lumber provided for promotion of the war effort
substantiated the old adage that 'It's an ill wind that bloweth no man
good.'"

EIGHT

The New Forest

If a tree falls in the forest and nobody is there to hear it, does it make a sound?

Take away the person who could hear the tree fall, remove humans from the picture, and the billion trees that Thirty-Eight knocked down would have been purely an ecological event. The cataclysm that surrounded Fred Hunt beneath his protector pine would have been simply a rearrangement of forest structure, increasing its complexity at ground level. The pine needles, twigs, and branches that Ward Shepard feared as a fuel load would have provided a burst of nutrients, especially nitrogen, to the forest soil. The prostrate sugar maples that reduced the Worthleys' and the Whites' spring cash flow would only have excavated pits and mounds in the forest floor, bringing mineral soil to the surface and burying organic layers.

Thirty-Eight was different from the hurricane of 1635 and those that preceded it because the stricken area in 1938 was home to nine million people. When the trees fell, people heard them. And that makes it difficult to assess the ecological consequences of Thirty-

Eight. People couldn't stand by and let their vertical world remain horizontal. We had to right the wrong, and we did a bang-up job cleaning up the mess. Harvard Forest's David Foster is not alone in thinking that the full-on human effort produced more change to the system than did the wholesale rearranging of the temperate forest by the tropical cyclone. On a walk in the woods in Petersham, Foster told me, "It's difficult to say what the effects of Thirty-Eight were since it's so confused in our minds by all the salvage that happened after. Thirty-Eight was a phenomenally destructive event that was followed by a remarkably efficient and comprehensive cleanup activity."

We have good documentation of the extent of the disturbance. The gaps it ripped in the canopy varied tremendously in size and shape. In one afternoon, 4 percent of the landscape was transformed from a mature forest with a relatively full canopy to one with almost all of its biomass concentrated within ten feet of the ground.

In any year, we can expect natural mortality for 1 percent of the forest trees, and we can assume that the rest of 1938 saw all of the standard mixture of natural disturbances. In any woodlot, insects and diseases killed a tree here and a tree there. Wind and fire took trees in larger patches, with wind knocking trees down a half acre or so at a time and the average fire consuming less than ten acres. And then in a span of five hours, Thirty-Eight took four times the expected annual toll. Beyond the areas of total blowdown, considerably more acreage suffered lesser damage that ranged from crown breakage to early defoliation.

If we had had satellites providing us images the day after the hurricane, the array of intermittent gaps of various sizes in the canopy would have definitely showed up, though the view might not have seemed as catastrophic as it did to those staring out their kitchen window at chaos. From above, the gaps would still have been green. To be sure, many leaves would have been blown off, and they'd be drying out and releasing nutrients, but most trees would still be alive because of the hinge of root that maintained contact with the soil.

If we could see an image taken in midsummer of the following year, a clear distinction would come into focus. Rather than a duality in black or white, the difference would show as green or brown. We'd see green in the downed hardwood stands, but the pine stands would more likely have been rendered brown. Pine dies rapidly when blown over, the needles drying and falling off. It can't sprout as hardwoods routinely do when cut or otherwise damaged, sending out new shoots from the stem, stump, and roots. More important, the CCC, WPA, and other crews had spent the fall and winter stripping flammable material from the downed pines. These defoliated areas were disproportionately in Massachusetts, where white pine dominated the blowdown and the populace was particularly wary of fire. There, the effect of the salvage operation compares to that of a contemporary whole tree harvest. A swarm of workers disposed of trees as thoroughly as a team of today's feller-bunchers and grapple skidders. Even if NETSA's high standards for logs meant that some stayed in the woods unsold, fire hazard reduction dictated that the trees were stripped bare. These areas took longer to green up and lost the nutrient benefit of a year's worth of needle drop. The ash in the burn pile remnants would act as fertilizer, but all the organic matter in the soil beneath the pile might have been incinerated depending on how hot the brushpile burned.

Fuel reduction in the northern hardwoods of Vermont and northern New Hampshire was much less thorough. This was not only a matter of the hardwoods being less flammable; up north, even the fallen pine wasn't routinely picked clean. Almost half of the man-days devoted to cleaning up the woods were spent in Massachusetts alone. With the addition of Connecticut and Rhode Island, 60 percent of the 215,000 acres undergoing intensive fire reduction was in

This roadside section of woods was salvaged for pulp and the fire hazard completely removed. The toppled forest was adjacent to the Franconia-Littleton Road in Grafton County, New Hampshire. United States Forest Service

southern New England. New Hampshire salvaged 58 percent of the logs but cleared only 21 percent of the acreage. The Blodgetts' and other northern pine stands were salvaged, but the brush remained in place. In keeping with the utilization standards of the day for both hardwoods and pine, loggers removed only logs and left the limbs and tops where the tree fell. Firewood gleaners might go back in later and grab undersized hardwood logs and limbs they could reach by truck, but a surprising amount of biomass was left in the woods. For ecologists, forest structure begins at the ground level, and dead wood plays an important ecological role. Each rotting trunk houses a web of animal life, with its life and death processes enriching and moistening the soil through decaying wood. As they decompose, the trunks shelter amphibians and small mammals from predators. Where pine and the fire threat coincided, little of this forest structure remained intact since eager crews removed much of the biomass.

Another ecological consequence that can be attributed to the cleanup effort rather than to the blowdown was the temporary increased flow of water. Herb Bormann and Gene Likens, in their influential book *Pattern and Process in a Forested Landscape*, provide a great picture of forest hydrology: "Precipitation is first intercepted by the canopy, then by litter on the ground. It is channeled by soil structure through the ground rather than over the ground, and before streamflow from the ecosystem can occur, hydrological storage capacity must be satisfied. Water enters the forested ecosystem with the potential of a lion and, most often, leaves meek as a mouse, with much of its potential energy lost in small frictional increments or simply by conversion of the liquid to vapor by evapotranspiration."

Following the cleanup, with considerably less foliage intercepting rain and transpiring it back to the atmosphere, fewer prostrate trees releafing on their way to eventual death, and less leaf litter on the ground to absorb it, more water flowed over the ground to streams. A Forest Service researcher in the 1970s on temporary assignment at Harvard Forest studied posthurricane flows in the Connecticut and Merrimack rivers, both of which drained areas devastated by the hurricane. The rivers showed significant increases in the flow the summer following the hurricane, and it took five years for the flows to return to their long-term average. The researcher compared these flows to those in the Androscoggin River, whose watershed largely escaped damage and predictably showed no increase.

No studies were made of soil erosion from the logging operations, but we can make some assumptions. The skidding technology of the time was relatively light on the ground. Horses and oxen pulled logs a short distance to the skid roads. There, the logger loaded them on scoots. Instead of skidding a hitch of logs with the trailing end dragging on the ground, crawler tractors pulling scoots kept the load from plowing up the ground. Unlike the rubber-tired skidders that came along in later decades, crawlers distribute their weight evenly across the surface where the tracks travel, so they are less prone to creating deep ruts. Still, on slopes, skid trails could definitely become streambeds. In the areas of intense cleanup, more of the leaf litter was disturbed and the soil churned up, and these soil disruptions added to that portion of the forest floor overturned with the upended trees. This stirred-up soil was a tailor-made seedbed for the small wind-scattered seeds of early successional species, such as birches.

And that, of course, was the major ecological consequence of the hurricane: mature timber was replaced by seedlings and sprouts competing with forbs and shrubs. In salvaged areas, the deep shade of a closed canopy became a stark, bright patch of brambles you'd avoid on a summer day unless the raspberries were ripe. This transformation occurred in thousands of canopy gaps across the landscape, some in small sizes, some quite large. The magnitude of this sudden change was enormous, but because of the intermittent and varied nature of the blowdown, ecologists have come to consider it not a single major disturbance but thousands of simultaneous minor disturbances. We tend to think of large disturbances as consuming everything top to bottom over broad areas. But that's not the way it works even in the large western wildfires, where some stands burn completely but others retain crown cover. The heterogeneity is even more pronounced with hurricanes. Thirty-Eight destroyed thousands of vulnerable pockets within a wide swath of New England. The heterogeneity of the disturbance was compounded by the differences in the degree of human activity that followed.

David Foster was so interested in studying the effects of an un-salvaged hurricane that he and his colleagues simulated one of their own in October 1990 using a skidder's winch to pull down 80 percent of the canopy trees in a two-acre study plot in Petersham. The skidder did its work from outside the perimeter, and all the toppled trees remained in place, so the stand has redeveloped on its own. They chose a plot typical of the central New England forest cover today, a mix of red oak, red maple, and associated species including birches and pine. In monitoring it and nearby control plots over the years, they learned that the pulldown had remarkably little impact on key ecosystem functions. The soil did not warm up, soil moisture didn't increase, and soil respiration didn't change. Organic matter decomposed at normal rates, and nitrogen cycling changed only modestly. In the years following the pulldown, the species mix in the two-acre patch of woods has remained relatively constant. The Harvard Forest ecologists wrote, "In the face of apparent forest destruction, many fundamental processes were maintained. Natural hurricane disturbance is actually not highly disruptive to ecosystem integrity." They attribute this stability in the midst of chaos to the fact that the vegetative cover was maintained even though the canopy had been dropped from nearly one hundred feet to ten feet.

The nature of the forest that began to grow in Thirty-Eight's wake ultimately depended on a number of factors, including the makeup of the toppled forest, how completely it had been knocked down, the size and shape of the gap, and whether it had been salvaged and cleaned up.

In ecological terms, disturbance and forest succession were battling over control of the future stand. Ecologists define forest succession as the generalized pattern of changes in species composition over time as one assemblage of plants outcompetes and succeeds another. As long as natural or human disturbance doesn't intervene, succession proceeds in a somewhat predictable fashion. starting with species intolerant of shade and moving on to more tolerant species.

Presented with a wide-open piece of ground by blowdown, logging, or pasture abandonment, pioneer species are well suited for rapid colonization: most produce plentiful, wind-borne seeds, and under full light and with abundant water and nutrients, they can grow exceptionally fast as seedlings. These sprinters have a competitive advantage over other species that don't spread so easily or grow as quickly as seedlings.

Over time, conditions change. As they mature, the intolerant species help create shady conditions in which they have to cede the ground to shade-tolerant species. Early-successional pioneers such as paper birch and quaking aspen cannot compete in the dim light of the understory, where sugar maple and other shade-tolerant species grow well, albeit slowly. The late-successional species last longer and eventually outcompete and outlast the pioneers.

If these forests escape major disturbance long enough, late-successional species rule the ground. Sugar maple, American beech, and eastern hemlock are built for a marathon and can maintain control a long time—hundreds of years—without being replaced by other mixes of species. That's because these species' shade tolerance means that their seedlings and saplings can survive in the understory much longer than white pine or other less tolerant species that will wither and die. Not only can they survive, but the tolerant species in the understory can thrive when given a flood of light after decades in the dark. New generations of the same species replace their parents, so the mix of species remains relatively constant for centuries. This is what ecologists once called a climax forest. It can take hold only in the continued absence of major damage from fire, wind, or logging.

Disturbance can interrupt this general pattern at any time because it creates available growing space by eliminating the plants previously occupying the space. Think back to Thoreau's address on

TABLE 2

SHADE TOLERANCE OF COMMON NORTHEASTERN TREES

Tolerant	Intermediate	Intolerant
Softwoods		
Atlantic white cedar	Eastern white pine	Eastern red cedar
Balsam fir		Tamarack
Eastern hemlock		Red pine
Northern white cedar		Pitch pine
Spruce (white, red, and black)		
Hardwoods		
American beech	American chestnut	Aspen (big-tooth and quaking)
American hornbeam	American elm	Birch, paper
Basswood	Ash (white, green, and black)	Birch, gray
Boxelder	Birch, black	Black locust
Eastern hophornbeam	Birch, yellow	Cherry, pin
Maple, sugar	Hackberry	Cherry, black
	Hickories	Sassafras
	Maple, red	Willows
	Oak (white, black and northern red)	Yellow poplar

Note: The United States Forest Service designates red maple a shade-tolerant species. Many foresters in New England consider it to have intermediate shade tolerance.

the succession of forest trees, and how the nearly invisible oak in Concord's pine stands would take over the stand after the pine was cut. Oak is a later-successional species than pine. In Thoreau's example, the natural, inevitable forest succession from pine to oak was accelerated by the human disturbance of the logging. Otherwise, the pine could have held on considerably longer until the death of individuals or small groups opened up the canopy and allowed the oak saplings or sprouts to thrive.

181

Thoreau could just as well have been talking about Thirty-Eight, with the hurricane replacing the logging as the catalyst. In pine stands that had hardwood seedlings and saplings waiting in the understory, those hardwoods became the next forest. When the pine overstory was removed, the hardwoods that had seeded in beneath the pine—the advance regeneration—had a leg up on everything else.

Salvage logging further encouraged hardwood's ascendancy over the pine. Two graduate students at Harvard Forest, Ralph Brake and Howard Post, documented salvage harvests in Petersham and the regeneration that resulted. "Only a few advance growth stems were left uninjured following logging. A great majority were cut back to the ground because they interfered with cutting and extraction of logs. This practice, though unintentional, has resulted in many rapid growing sprouts," they wrote. They saw sprouts of red maple, gray birch, black cherry, and red oak quickly take over many stands, outpacing seedlings because sprouts can draw on the food reserves of an already well-established root system, allowing them to concentrate growth in their stems.

This pattern of sprout and seedling hardwoods replacing white pine prevailed in 80 percent of the white pine stands. The hurricane greatly reduced the presence of white pine, though the exceptions to this pattern are notable. Some soils just want to grow pine, and where it had the benefit of a good seed year, pine maintained itself in the species mix, though not always in pure stands. Playing his part in the sawing of all the downed white pine in the Merrimack valley, Joe Colby said to his son, "Jim, there won't be a stick of pine timber left." But afterward, he kept sawing in Boscawen, Canterbury, Webster, and Salisbury, and he never ran out of pine lots. His son Jim said, "I started in business for myself in 1947, and I'm still not going out of any of these towns. You still find pine timber. It all depends on the land. On good, decent land, pine will seed fast."

Decent land to Jim Colby, whose livelihood has always depended on white pine, has light, sandy soils where pine competes well because hardwoods find them inhospitable. These river valleys were growing pine when the settlers arrived, as were glacier-influenced landforms such as outwash plains and kames. On heavier soils, the new forest also reverted to an approximation of the presettlement forest, but there it was hardwoods and a mix of hemlock, pine, and hardwood.

Let's take a closer look at what was happening in those stands now mostly void of large standing trees. The circumstances of each stand differed, but four sources of new trees were available to a varying extent. Some seeds that lay buried in the forest duff were triggered to germinate by the increased light or soil temperature. Pin cherry seeds, for instance, may have been buried for many decades, having been deposited there in the wake of a previous disturbance. Other seeds were blown into the opening by the prevailing winds. You may be familiar with the helicopter-style seeds of maples, called samaras. You may even have done what we did as kids, separate the flaps at the base and attach it to your nose. The wings have a purpose other than amusing children. Many species attach wings or other flight enhancers to their seeds so they can disperse far from the parent tree. Birch, pine, and aspen have very light seeds that can travel far on the breeze, the cottony aspen the long-distance champ with a range of several miles. Life as a seed is tough, and chances of survival are infinitesimally small, but not quite zero, which is why trees produce such an abundance of them. One study showed that five and a half million viable sugar maple seeds per acre were produced in one year, and less than 7 percent of them survived for a year.

Two other sources supplement buried or new seed. Seedlings or saplings that survived the crashing of the overstory and the subsequent salvage would be well positioned as advance regeneration to take advantage of the flood of light as long as they weren't badly shocked by the starkly altered conditions. Sprouts from hardwood trees and saplings cut at ground level in the cleanup would have the reserves of an intact root system to draw from, a nice advantage.

In any one piece of ground, one strategy would be more successful than in others, but the possibility of all four strategies means that the diversity of species would have been staggering in those first few years, providing the richest diversity of any stage of the development of that stand. Similarly, many animals would have flocked to the abundant food and cover conveniently brought down so close to the ground. Depending on local circumstances, the first cohort of trees might take as long as a couple of decades to establish itself and eliminate further invasion by others. Regardless of the duration of its establishment, foresters refer to the new forest as an even-aged stand. Chad Oliver and Bruce Larson, whose book *Forest Stand Dynamics* has been influential in forestry circles, prefer the term *single cohort stand* to *even-aged*. Both terms mean that the group of trees growing on the site all had their start following the same disturbance. Because trees can regenerate from any of the four strategies—buried seed, new seed, saplings, or sprouts—a tremendous variety of species can be present, including late-successional trees. Researchers at Harvard Forest documented that the species that controlled sites forty years after the hurricane—and would control it for many decades in the future—were in place in the first ten years, with many of them sprouts from fallen or cut trees.

This flies in the face of the classic view of forest succession, that one species follows another in a sort of relay, which gave rise to the term *relay floristics*. A set of species holds the ground for a certain length of time, thus preparing the way for the following set of species. Ecologist Frank Egler coined another phrase, *initial floristics*, for the opposite explanation, that trees of many species, including late-successional species, invade a disturbed site all at once. Combative and iconoclastic, Egler was so certain that initial floristics controlled forest succession that he offered a $10,000 prize to anyone who could confirm an instance when relay floristics occurred. He made the challenge in the 1950s and republished it in 1975, and no one ever submitted a claim for the prize.

William Drury, a promoter of Egler's ideas and himself a brilliant ecologist, pondered how botanists could have ignored initial floristics. He concluded that by focusing on the flood of early-successional species that control the site, they were blind to any other species in the mix. Like Thoreau's oak seedlings, they were not immediately apparent. Making his own case against the relay theory, Drury pointed out that natural selection would hardly favor species that "foul their own nest, creating an environment where they can no longer persist." In other words, it wouldn't make evolutionary sense for aspen to compromise its own prospects and prepare the way for other species.

Drury had studied with Hugh Raup, one of the first heretics to question the succession orthodoxy. Raup was a botanist who made his name at the Arnold Arboretum, at Harvard Forest, and in the frozen north of Alaska, Canada, and Greenland. His quarrel with climax began in the 1930s (he was born in 1901), and he later articulated those doubts in a 1972 lecture on forest history he gave at Rutgers University. Raup noted that the idea of succession to climax had come from two prominent ecologists, Henry Cowles and Frederick Clements, who made it "the central theme of American ecology. It was the ecology and the ecological plant geography that I was taught when I went to college and graduate school," he said. "It is now so deeply entrenched in our educational programs that children get it from the first grade on."

Emboldened by the work of Henry Gleason, who first questioned Clements's views, Raup dismissed the theory, attributing its persistence to its readily understandable logic and its own ready-made jargon of words "that sounded as though they meant something." Raup wondered aloud how ecologists who at that point had been theorizing about succession for only six or seven decades could describe forest successions that played out over centuries. The only way to confirm their truth was to "sit with a notebook and watch for a century or so. My personal experience leads me to think that most of the successions were spurious. I know that mine are."

In his talk, Raup plunked succession, then took aim at another outworn concept, climax. The climax stage had certain prerequisites, especially the exclusive presence of late-successional species. These species must be present in all ages, ranging from seedling to old age. The only way they could persist was by living through a long period with minimal disturbance. He wrote, "It had to be long enough to cover the life spans of not only one, but of several generations of trees. In our northeastern American forests, as well as in many others, this time period would easily extend to a millennium."

Raup refuted these possibilities by pointing to the history of the forest in Petersham. A 1793 account described that part of the town's forestland that had not yet been cleared for agriculture. The swamps and lowlands contained species one would expect of a climax forest: beech, birch, maple, and hemlock. But on the high lands, the chronicler said the growth was "oak, more chestnut, and a great deal of walnut of later years." By walnut, he meant hickory, a backbone of that forest. None of these nut-producing species—oak, chestnut, hickory—are shade tolerant or late successional. They would not dominate anyone's picture of a climax forest. Thus the forest in place when the region's settlers arrived was not a climax forest.

If not, why not? Because succession had been interrupted by a disturbance, probably a hurricane but possibly fire, that reset the successional clock. There could be no inevitable climax condition because disturbance is so frequent and inevitable.

Raup and other disturbance ecologists had confirmed that hur-

ricanes were prevalent in the region for perhaps as long as trees
have grown there. He concluded, "The supposed climax forest of
tolerant trees was not in our region when the first white settlers
came to it, and now we cannot believe it was ever there. One of the
basic requirements for its development—hundreds of years with no
serious disturbance—probably has never been realized. Further, the
primal forest being made up of all ages also goes by the boards, because a disturbance as large as the hurricane produces whole stands
of forests with trees the same age."

In Raup's formulation, which has been widely adopted by forest
ecologists, disturbances of all sorts are the norm, and forest succession begins again after disturbance hits the reset button. It's as if an
actor is trying to get all the way through a scripted speech from beginning to end but the director lets him go only so far before calling
out "Cut!"

The nature of scientific research consists of verifying or correcting assumptions about the way things work. Scientists spend their
careers disagreeing with predecessors or colleagues, but they dispute while following a certain decorum. Academic papers are invariably written in the passive voice: "It was determined that . . ." Nowhere
will you find the word *I*. That's why Egler's audacious $10,000 chal-
lenge was so startling. Raup likewise seems supercharged by his
animus against succession. He was not correcting an error, he was
righting a wrong.

Indeed, there seems to be a bit of an ontological battle taking
place, as climax suggests a level of trust exasperating to an agnos-
tic. I can remember first learning about a climax forest as a teen,
hearing that designation given to the woods in the Adirondacks I
was so familiar with. I understood it to mean a forest made up of
big, old trees, a forest that had reached an immutable state of splen-
dor. The trees were huge, the shade was deep, and the smells were
luscious, moist, and earthy. These were woods as they were meant

to be. For someone who'd been educated by nuns and spent most Sunday mornings as an altar boy serving Mass at the neighborhood church, the climax forest was a comforting notion akin to absolution and plenary indulgences. It was nice to know that there was a system in place, that someone was in charge, and that life happened in an orderly fashion. A climax forest seems to assure that "God's in his heaven, all's right with the world."

Proponents took it so far as to maintain that a climax forest is a superorganism, a holistic community. The species that form the community have coevolved and they depend on one another. If one member of the community disappears, the whole is at risk. Utter nonsense, say Egler, Drury, Raup, and a host of others. They sputter as they make their case that an individual plant grows in a place simply because it finds the conditions suitable. Other species grow alongside for the same reason. An assemblage does not imply anything more than that each species individually competes well on that ground. Remove one species and its absence will cause not an ecological swoon but an expansion by the others or the introduction of a new species. The balance of nature is wishful thinking. There is no steady state, no inherent equilibrium that we will always return to. Period.

Their insistence that disturbance trumps succession is borne out in central and southern New England, with its long history of fire and catastrophic hurricanes that arrive every century or two. Hurricanes regularly set back forests in the southeastern United States, while fire has been an important player in all forests west of the Mississippi. Closer to home, fire can be expected to return to Lake States' forests every century. Quebec and the Maritime forests have similarly frequent experiences of catastrophic fire. The disturbance theory holds true wherever major disturbances occur frequently enough to reset the clock, which begs the question: Is there any area that escapes stand-replacing events for enough centuries to develop a climax forest?

One answer comes from Herb Bormann and Gene Likens, whose description of hydrology I quoted earlier, and who devoted their careers to studying the watersheds at Hubbard Brook Experimental Forest, in New Hampshire's White Mountains. In their work describing how forests develop over time, they defended the possibility of a steady state and made the case that if ever it could happen, the northern hardwoods of humid, inland northern New England would be the place. They wrote, "Within the northern hardwood forests of the White Mountains there is very little evidence, vegetational or historical, that fire was widespread. . . . After 1869, with the introduction of large-scale logging in the White Mountains, the frequency of fires increased drastically. For the most part these fires burned over slash-filled areas left by logging operations. Since the advent of fire protection, fire is once again a relatively rare occurrence." The widespread acceptance of this attitude toward fire has led to our northern hardwoods being known as the "asbestos forest."

Nor have hurricanes been a frequent part of their history. Bormann and Likens note that the White Mountains have experienced only two hurricanes—in 1815 and 1938—in the five centuries since Columbus. Most of Vermont has been even less vulnerable to hurricanes, since meteorological accounts show the 1815 storm track was far enough east that Vermont's damage was primarily flooding from rain, just as with Irene in 2011. Bormann and Likens didn't lump Vermont's northern hardwoods in with New Hampshire's, but at least in terms of overall disturbance regimes, we can think of them as one.

Bormann and Likens acknowledge Raup's contribution in refuting an outdated notion of succession and replacing it with disturbance as the primary factor in determining what plants live where. But they suggest that the pendulum of ecological thought has swung too far in that direction. Disturbances are important, but they're not equally important everywhere. They proposed that a steady state *could* exist if viewed as what they termed a "shifting mosaic." Within the broad swath of northern hardwood forest that is progressing toward an older condition akin to a climax forest, there are gaps in the canopy formed every year where younger forests reestablish themselves in the wake of smaller-scale disturbances. Only if the disturbance and the resulting gap is large enough can early-successional species take hold. The space created by the gradual death of an individual tree will be filled by the lateral expansion of neighboring trees before any new trees can grow into the canopy.

Vermont and New Hampshire's northern hardwoods—the birch, maple, and beech forest that draws a parade of foliage tourists—are prone to frequent small-scale disturbances. Thirty-Eight, the one hurricane to hammer both states with 100 mile per hour winds, pummeled exposed sites and left protected sites alone. And even on the most vulnerable hills facing east and southeast, the results were often more like a bad haircut than a clean shave. Not all the trees were bowled over. At Hubbard Brook, on hills facing southeast, the damage was severe in some spots, with one area having lost 82 percent of its cover. Over the larger affected landscape, however, the canopy loss averaged out to 20 percent, largely because recent logging had reduced the age and size of much of the forest. All told, this means that in the northern forest, disturbance and succession played out differently from in pine country. Thirty-Eight knocked down thousands of northern hardwood stands at varying scales, and in those gaps new younger forests developed. Given the late-successional characteristics of beech and sugar maple and their capacity to nurture an understory of their own progeny, the species mix didn't change dramatically despite the loss of the overstory.

Only where the gap was an acre or more, and all the overstory trees were either blown down or cut as part of the salvage effort, could the early-successional species dominate the new forest. Light wind-borne seeds of paper birch, aspen, and white pine could have

The damage to this stand of Vermont hardwoods shows the characteristic patchiness of the blowdown. Shorter, smaller-diameter trees remain standing, maintaining control of the site. Vermont State Archives and Record Administration

seeded and taken control of these open sites. Even there, the pioneers would be competing with sprouts and seedlings that had been twiddling their thumbs in the understory, though they wouldn't also have to contend with mature trees shading them.

The Hubbard Brook scientists have been careful to portray their shifting mosaic steady state as a model and not an actuality. They define steady state in ecological terms that take into consideration more than the age and size of the dominant trees. Their steady state is an ongoing condition with no net change in total biomass as measured in four compartments: green plants, dead wood, forest floor, organic matter in the mineral soil. Their forest growth model goes through four stages culminating in the steady state. To reach this state of equilibrium, the forest will have to develop undisturbed for three or four centuries. They acknowledge that's a tall order, but they make it clear that human disturbance—not fire or wind—is the most likely agent of large-scale change. Before humans became a disturbing agent every bit as powerful as wind and fire, the northern hardwoods could have spent centuries in this shifting mosaic state.

Today, since logging can be expected to alter most stands of trees once they reach a certain size, it's much less likely to happen. Half a million individuals and families make decisions about the forests they own in New England, all of them with different needs and interests. Responding to survey questions, most forest owners profess little interest in harvesting trees since they own their woods for privacy, recreation, and other nondisruptive reasons. Harvesting records show the opposite, that even the most disinclined forest owners will bring in a logger when the timing is right and a persuasive financial offer is made. All of these episodic timber sales contribute to a different kind of shifting mosaic, one in which most trees don't reach their biological maturity.

Pecking away in this way bears no resemblance to the wholesale land clearing accomplished by Europeans soon after arrival, nor does it match the logging intensity of the pine boom that peaked in 1907, when 1.7 billion board feet was cut within the area later to be affected by Thirty-Eight. With clear-cutting out of societal favor, cuts today are lighter. Still, these harvests disturb forest stands bit by bit. Meanwhile, ice and snow, microbursts, and insects and diseases take their annual 1 percent toll.

New England's forest is growing, with annual growth exceeding harvest by wide margins except in Maine. Only in large reserves where logging is forbidden can forests be expected to reach anything approaching a steady state. Even then, the natural but exceedingly rare fire or hurricane or the increasingly common ice storm can reset the clock at some point. Still, there's no reason to suppose that the area couldn't remain free of major disturbance again for the next thousand years. As Hugh Raup suggests, none of us will live long enough to sit around with a notebook and document where forest succession leads. Only our own successors hundreds of years from now will know whether succession or disturbance controls these reserves.

NINE

Telltale Signs

Charlie Cogbill describes himself as a plant ecologist, but one of the first items he jotted down in his notebook on a visit to my woods was a list of the birds he heard singing: hermit thrush, black-throated green warbler, red-eyed vireo, all of them songbirds of deep woods. Later, as we neared the neighboring property where there had been a recent raggedy logging job, a chestnut-sided warbler sang *Pleased, pleased pleased to meetcha*, and his eyes lit up. Out came the notebook: different forest, different birds. Cogbill solves forest puzzles, and like any good sleuth, he takes note of all potential clues.

We were there to talk about plants, big plants, maple trees that had been gone for seventy-five years. I had sought out his opinion of some puzzling signs of hurricane damage in my woods, and he was intrigued. His curiosity and brimming enthusiasm for what he sees and hears in the woods makes him a great companion to tromp around with. That's why I've always found it so startling that much of his contribution to forest ecology has come from decades spent indoors and alone.

In libraries, archives, and town halls, he deciphers deeds and surveys written in seventeenth- and eighteenth-century hands, on the prowl for references to trees. Of his work, he said, "I've been lugging away at what I call the parallel tracks, one of which is working with the actual forests and their dynamics, and the other is working with the historical records and what they are saying about the composition and the dynamics of forest in the past."

In this way, he has helped define the forest that was here when the settlers recorded their deeds and started clearing it for farms. He told me, "The lots were laid out on plats ahead of time in somebody's tavern or living room. The surveyor then went out with a compass and chain, came to a corner, and typically put in either a post or a stake or a heap of stones. Then he made note of a nearby tree to reference it." These witness trees confirmed the monument in case of any boundary dispute but also give testimony centuries later to the species of trees present at the time.

Cogbill enters these witness trees by species into what is by now a huge database and through statistical analysis creates a picture of the presettlement forest town by town. Cogbill began this work at Hubbard Brook in the 1970s, and he's intimately familiar with the watersheds where Herb Bormann and Gene Likens measured streamflow, biomass, and soil chemistry for their model of forest development. Among the details his parallel tracks show is that Hubbard Brook is a primary forest—it was never cleared for agriculture. Since settlement, it has changed largely through two major disturbances—logging for spruce in the 1910s and the 1938 hurricane—and countless other smaller events.

His fieldwork has made him an aficionado of blowdowns, and seventy-five years after Thirty-Eight blew through, he could point out signs of its passing. The most common sign is the undulating ground where tip-ups (uprooted trees) transformed the forest floor. The ground in hurricane woods is pockmarked with deep depressions adjacent to correspondingly large mounds. Known as pit-and-mound topography, it looks as if somebody dug a hole and piled the dirt next to it, and it happens anytime the wind uproots a tree. The rootball of a forest tree forms a plate, and when wind knocks the tree over, the roots are wrenched out of the ground with only a hinge of roots maintaining its attachment to the ground. As the rest of the roots rip free, they grasp a mass of soil and stone, excavating a hole. The bigger the tree, the bigger the pit.

Over time, the roots and stump decompose, leaving a mound of earth. The pit collects leaves yearly, but they decompose without filling in the hole to any great extent. Decades after the tree went down, you can stand in the pit, look out across the mound, and that's the direction that the tree fell. In most tip-ups, the fatal gust of wind catches the tree's sail so it keels over in line with the wind direction.

In any locale east of Thirty-Eight's storm track, the first strong winds to strike would have been from the east; then as the center progressed northward, the winds would have been from the southeast and finally from the south. Particularly vulnerable trees blew down in the earliest gusts from the east, and they would have been laid down with the crown pointing west. Wind velocity increased as the storm closed in, and in most locations the peak winds were from the southeast. Trees that withstood the early east winds but ultimately lost their grip probably would have fallen pointing to the northwest. Most of the trees fell within a range of 290 to 340 degrees.

If you come across a half acre or more with pits and mounds pointing to the west or northwest, it's a sure sign that Thirty-Eight did the deed. A southeast wind strong enough to uproot trees is a rarity in interior New England, where the everyday prevailing wind tends to come from the west, so most gusts—including those from microbursts embedded in thunderstorms—blow trees down to the east. The layout of the pits and mounds will confirm that the wind came from the west.

If a remnant stump shows up in the mound, you can count on its

197

This pit and mound dates from the 1938 hurricane. The tree growing on the front of the mound established itself after the original tree blew down. The pit to the right of the mound is still deep nearly eighty years later. The dead tree beyond the pit fell recently. Stephen Long

being white pine, not just because that was the most common victim but because its wood persists so long. A pine stump has a distinctive annual whorl of branches, and these branch stubs are the hardest, most rot-resistant part of the tree, outlasting the trunk by decades. Yes, it's counterintuitive, but hardwoods, with the exception of black locust, disappear much more quickly. Cogbill says, "I use fifty years as my decay window for hardwoods. Everything over that you're looking at soil morphology, not looking at trees."

The mound might have a paper birch or yellow birch growing on it, and if so, the tree might be standing on stilts. The mineral soil of a tip-up is a perfect seedbed for the birch to establish itself immediately, and over the years as the soil sloughs off, the roots extend downward to maintain contact with the soil, becoming what's known as prop roots. Another clue: paper birch wouldn't have reached maturity in the shade, so its continued presence on a mound indicates that the blowdown that spawned it was not that of a single tree but was extensive enough to open up the canopy around it.

The form of a tree's trunk can link it to the hurricane. "Pistol butt" is one of many Cogbill coinages, and it refers to a tree with a slight kink in the first couple of feet above the stump that shows it has made a course correction. A pistol butt happens when a tree is tilted sharply and then resprouts from the lower trunk rather than the stump. The original trunk dies off, and as the sprout adds wood over the years, it can't quite hide its origins as a trunk sprout.

Saplings and poles don't uproot because they haven't built up the annual layers of wood that give fully grown trees a lever stiff enough to wrench their roots out of the ground. They can, however, hitch a ride down with a larger tree if they are growing within the reach of its roots. The sapling roots themselves haven't been broken, so they aren't at all compromised. Nor is the crown. What a great opportunity: the canopy trees that were shading it all these years are suddenly prone; light, the life force for trees, floods in. The young tree should be primed to take advantage of it, but it's growing in the wrong direction.

The larger companion tree might leaf out for a few years, but its root system has so little attachment to the ground that it will weaken and die. If the sapling isn't damaged by the possible salvage, it will do whatever it can to right itself. Phototropism—the ability in trees to sense sunlight and grow toward it—corrects the trunk's lateral lean and sends it skyward in either of two ways. If it's a de-

Each of these trees was bent at an early age. The hemlock on the left responded by recurving its trunk to an upright position. The elbow in the hardwood on the right (an eastern hophornbeam) shows that a side branch took over as the leader and the original leader rotted away. Stephen Long

ciduous tree and the angle is acute enough, one of the branches on the top side of the prone trunk will exhibit leadership skills and take over as the terminal leader, the dominant stem. Beautifully positioned to head vertically to the sky, this branch will take off, growing both in height and girth. The new leader will succeed so well that the tree will give up on the wood growing out beyond it. Over time, the former top will slough off. What you see decades later has a boomerang shape, or if the remnant section of original trunk is lying almost flat, the new section will be growing almost at a right angle to it.

A small hemlock faced with the same bent-over situation often will handle it differently. If the terminal bud is not damaged, the leader will not so readily cede its authority and will itself curve upward. This is known as epinastic control, a trait that is strong in hemlock and other conifers, which tend to stick with a single stem from ground to sky. Your childhood drawings of trees illustrate this concept well: conifers have the single-stem, Christmas-tree shape, and hardwoods look like lollypops. Indeed, most hardwoods' stems—short if grown in the yard and tall if forest grown—will fork into secondary trunks, limbs, and branches convoluted enough that it can be hard to identify the original leader. Consequently, when knocked out of plumb, softwoods and hardwoods tend to respond differently: monarchy for one, anarchy for the other.

A tree shaped like a boomerang or having a gently curved recovery form can happen anywhere in the forest, so the presence of one is not necessarily an indicator of hurricane damage. The direction it leans is the key. If its lower trunk leans somewhere between due north and due west, it means the wind came from the south or the east. Nearby pits and mounds in the same direction bolster the case that Thirty-Eight leaned on the tree.

What other indicators of hurricane damage might you see more than seven decades later? Older trees that lived through the blow might show that they sustained crown damage and survived by responding to it. After the leaves drop, it's easy to see irregularities in tree crowns. The break may be apparent, but sometimes you can see only the response to the break. The loss of a considerable part of the crown causes the tree to compensate with a spurt of new growth in the adjacent branches. A new leader can form from a shoot growing from a horizontal branch. Or it will grow new branches—called epicormic branches—from dormant buds on the surface. The stem may swell at the point of all the new growth almost like a graft on an apple tree. Two or more branches can compete to replace the broken leader, and each will display a crook as it gravitates toward plumb. This kink looks similar to a pine's response to a weevil.

Any of these crown repairs should probably be considered as secondary evidence of hurricane damage, because upper limbs can break in so many other ways. Snow loading on leafed-out trees in spring or fall can snap limbs, as can a buildup of ice in the winter. Crowns can break at weak forks from any strong wind. If the old deformed crown is found in the company of nearby pits and mounds or leaning trees, the damage probably came from the hurricane.

I showed Charlie Cogbill some old trees in our woods, several sugar maples that grew along a property line with open pasture on both sides. They now have younger forest surrounding them, and we could surmise their pasture origins because their heavy lower limbs began at eye level, just beyond the reach of cows. Some of them were three feet through, a result of their adventitious position of a lifetime in full sun. Big, sprawly trees like these have no commercial value, and foresters once termed them *wolf trees*, predators that make life difficult for the young ones in their shadow that presumably could otherwise make something of themselves. Forest owners less obsessed with yield per acre enjoy seeing huge trees in their woods, prefer the term *legacy trees*, and resist the advice to cut them.

I showed him a stand on a steeper slope that was never cleared for pasture, where another old sugar maple has an entirely different history. The dominant tree in its neighborhood, it forks twenty feet up into a spreading crown. At twenty-eight inches across, it has twice the diameter of its neighbors. Some of the bark is sloughing off its lichen-covered trunk, but no woodpeckers have been probing it for insects, so it seems sound. Its elongated form confirms that it grew up competing in a cohort of trees. In the long process of self-pruning, it was shedding lower limbs before the Civil War. This maple was a seedling in 1832, the year Andrew Jackson was elected to a second term as president. Such longevity makes it worthy of a name, and I've christened it Andrew Jackson. Yes, this old maple is named after Old Hickory.

Loggers passed it by on various occasions, probably because it has a prominent seam and a pronounced lean that probably has caused compression wood, either of which would compromise its lumber value. It had a fifteen-inch diameter in 1955, when logger Walter Dunklee bought the timber rights on this lot and drove his Army surplus deuce-and-a-half deep into the woods to truck the logs out. He didn't cut it. Nor did he cut it a decade later when he returned as the land's new owner.

Today all the trees in its vicinity that don't compete directly with it for sun and nutrients have nearly caught up with its ninety-foot height. Potential height is determined largely by soil conditions, so trees of the same species that grow unimpeded on the same site can be expected to reach a certain height and no more. Better soils, taller trees. These soils are rich, but height growth can't continue indefinitely, and Andrew Jackson probably stopped its height growth forty years ago.

A tree's diameter growth, on the other hand, depends on how much light reaches its crown. A suppressed shade-tolerant tree can expand precious little while a dominant tree—with no adjacent trees encroaching on its crown—can add girth in a hurry. Andrew Jackson has experienced both rags and riches. In one four-decade-long suppression beginning in 1847, it added only an inch and a half in diameter. In a subsequent four-decade period after its crown suddenly gained its freedom, it celebrated by adding twelve inches in diameter.

The other canopy trees on this steep east-facing slope range in

This twenty-eight-inch-diameter sugar maple was established by 1832 and survived the 1938 hurricane, while many of the trees surrounding it were blown down. All the other trees in the photo have established themselves since then.
Stephen Long

diameter from twelve to sixteen inches. They form a single cohort with an average age of one hundred years. In 1938, they were all saplings, none of them having a diameter larger than three inches. All of these trees survived the hurricane.

I owe my knowledge of these trees' history to a forestry tool called an increment borer and to Dave Orwig, a forest ecologist at Harvard Forest who showed me how to use it. Orwig has cored thousands of trees, counting their annual rings to determine their age. More important to Orwig and other dendrochronologists, though, is what the tree's long-term pattern of growth tells them about the conditions it lived through.

An increment borer is a long, hollow, threaded drill bit. When it's muscled deep into a tree and aimed for the center, it envelops a core of wood the diameter of a pencil. You then insert a semicircular dipstick to withdraw the core from the tree and place it gingerly in a drinking straw for safe transport back home. After gluing the core into a groove cut in a narrow board, you sand it so that you can see the annual growth rings. Each year shows itself through a distinct line that separates it from the previous year. Orwig set me up at a microscope and showed me how to use dendrochronology software to record each annual increment of the tree's growth.

That autumn, we had a logging job in our woods, which provided me with other tree samples to study. With a chainsaw, I sliced cookies (inch-thick cross sections) from the butt ends of freshly cut trees, sanded them, and examined them under the microscope in the same way. With data from fifteen cookies and cores, Orwig showed me how to interpret what I found.

Counting backward from the bark, I reached 1938, which brought me almost to the pith. Along the way, there were variations in growth, but none showed growth as slow as that before 1938, when it had taken these saplings an average of fifteen years to add an inch of diameter. Clearly, they'd been stunted in the understory.

In reckoning their birthdate, Oliver and Larson would age all of these advance regeneration saplings from the time of their release in 1938, not from their emergence as seedlings. This approach makes great sense because they function as a single cohort. Still, I find it useful to know the specific prerelease history of these fifteen trees, because it shows what a formative experience the hurricane was for them. Nine of them started as seedlings between 1900 and 1920, which suggests that a disturbance around 1900 was large enough to give them a chance to establish themselves. Then when the overstory trees blew down in 1938, these smaller trees didn't blow down with them, nor were they crushed. Their final bit of good fortune was that they managed not to be in the way of the loggers salvaging the fallen timber. So when they were suddenly free to grow, they took advantage of it and grew rapidly for the next few decades. A few of them blasted right off, but others required two or three years to recover from the sudden physiological shock, at which point they too sped along. In the three decades following the hurricane, the average tree added five and a quarter inches in girth. A forest of saplings had become a forest of poles.

Andrew Jackson's experience differed only in that it was a larger tree when the hurricane hit. It had taken a century to grow eleven inches (hampered by that slow four-decade suppression), but following the hurricane it added nine inches in just thirty years. It nearly tripled its growth rate. Clearly, for this tree the hurricane was a godsend.

In *Forest Stand Dynamics*, Oliver and Larson differentiate between major and minor disturbances in terms of the new forest that results. Following a major disturbance, newly regenerating trees compete only with other regenerating trees, but when the disturbance is minor, newly regenerating trees compete not only with their peers but also with larger trees that survived the disturbance. In broad terms, major disturbances spawn even-aged or single-cohort forests, while minor disturbances promote or maintain an uneven-aged condition. Still, they note that the distinctions between the two can be blurry depending on "if the forest floor vegetation is well developed and survives the disturbance, or if only an occasional overstory tree remains."

Both of these pairs of seeming opposites—major-minor and even-uneven—function more as a continuum. It would be hard to dispute that sixty acres of blowdown and salvage is a major disturbance that results in an even-aged forest; similarly, a quarter-acre hole where mature trees expand their crowns and begrudgingly allow some advance regeneration qualifies as minor and promotes uneven conditions. Thirty-Eight spawned plenty of the latter, but the hurricane patches large enough for new trees to take over the ground have had far more lasting significance. As the army of foresters spread across the land and identified 600,000 acres of blowdown, it was the larger patches that made it onto their tally sheets.

Oliver and Larson explain how these forests developed through their stand dynamics model, which parallels Bormann and Likens's biomass accumulation model, each recognizing four stages. The silvicultural focus has forests progressing through stand initiation to stem exclusion to understory reinitiation to old growth. Stand initiation follows a disturbance, when all the seeds, sprouts, and seedlings/saplings produce a welter of green at the ground level. Each species employs its own strategies to establish a place for itself at the table, some of them shooting for the sky, others developing strong roots and photosynthesizing in low light. By the time World War II ended and the soldiers returned, 600,000 acres of young New England forests were denser than Times Square on V-J Day.

The next stage—stem exclusion—began soon thereafter, the canopy's height growth led by trees with dominant crowns. Stems are excluded in two ways. The complete shade on the forest floor stifles germination and precludes the growth of the rare successful germinants, so any new plants are excluded. Many of the competing new trees are excluded in what amounts to a single-elimination tournament in which a dominant individual vanquishes adjacent trees by expanding its roots and crown. Expansive root systems gain most of the soil nutrients and water, while the leaves in dominant crowns garner the available light. The dominants' lateral branches abrade the leaders of shorter trees like the bully who uses his superior reach to slap down those who challenge him. In this way, an acre that started with a million seedlings drops down to a thousand saplings, then half that number of poles. The lost trees starve because their suppressed roots and leaves can't keep them going.

As Egler showed with initial floristics, any species that provided parent material after the hurricane enters the competition. Sometimes the tortoise overtakes the hare. Mixed stands of red oak, red maple, and black birch abound in southern New England, and in many cases red maple took the early lead but couldn't keep up the pace, giving way to red oak after twenty years. In any place that pin cherry sprinted to a leading position in the canopy, it precociously spread its seeds for a decade or two but disappeared within thirty years. The adult tree died, but its seeds can stay viable in the duff for an extraordinarily long time.

The stem exclusion stage, with its characteristic full canopy and barren understory, can endure for many decades. An overstory of white pine or other intolerant species keeps excluding stems for fifty or sixty years, while the red oak–red maple–black birch complex holds its ground for eighty or ninety years. Northern hardwoods can continue for more than a century with an understory so sparse it looks like a city park, minus the lawn and drinking fountains, of course.

In Charlie Cogbill's study of Hubbard Brook, he and his colleagues tried to tease apart the history of the various stands. As in many northern hardwood stands, the hurricane damage had been patchy. Aerial photos taken in 1978 showed that the canopy had grown up and filled in so successfully it was difficult to discern from the forest structure what had or hadn't been damaged. He wrote that "40 years of canopy development effectively masked the patchy effects of hurricane damage."

Today, almost forty years farther along, the signs are even more obscure. Stands that grew back to white pine have already entered the next stage of development, known as understory reinitiation. As canopy trees grow taller and limbs longer, overlapping limbs rub against each other until the tips break and they no longer overlap. Trees that have deployed sharp elbows in the stem exclusion stage now maintain some elbow room around their crowns in a phenomenon known as crown shyness. All of these canopy trees have won their single-elimination tournaments, and nobody is organizing a tournament of champions to determine a single winner. On the cusp of the new stage, an acre of mature pine can hold one hundred or so large trees. Mature trees no longer fill in the available gaps as readily as before, so when a vanquished competitor—which itself would have to have a substantial crown to have made it this far—dies in place, a relatively large space becomes available for new growth. More light penetrates to lower crowns and to the forest floor.

Through the years, the forest floor has seen periodic attempts by germinating seeds to prosper as seedlings, and these efforts have heretofore been beaten back successfully. Finally, the canopy conditions change enough that sufficient light allows these seedlings to survive. A minor disturbance can open up the canopy prematurely and have the same effect. Only at this advanced age does a single-cohort stand admit a second cohort.

The stand of sugar maples and white ash that Charlie Cogbill and I were pondering was already in the understory reinitiation stage, the process having been accelerated by a forest thinning harvest a decade earlier. In response to the opening of the canopy, a flush of low green replaced the previously barren brown of the forest floor. The green, though, does little to mask the prominent pits and mounds scattered across the slope.

This minefield provided the puzzle I had asked Cogbill to help me solve. Facing east and southeast, this steep slope rises abruptly at the end of a long, broad meadow so it would have worn a bull's-eye for Thirty-Eight. Dendrochronology showed that the current crop of trees had been released from suppression right after the hurricane. Furthermore, the land records at the Corinth Town Hall showed that George Hastings, a neighboring farmer, purchased this parcel in April 1939 and promptly sold to the local bobbin mill the timber rights to all of its "beech, birch, and maple timber standing and blown down." The timber rights were to last for seven years, but evidently the bobbin mill's loggers got right after it because Hastings was able to resell the land in 1941.

There could be no question that the trees had been blown down in the hurricane, but there was a problem: the pits and mounds didn't face northwest. Instead they faced east, suggesting the culprit was the west wind. And because west is the prevailing wind direction, the trees that excavated the pits could have been blown over in smaller storms at other times. For instance, the microburst I described in Chapter 3 hit us from the west and laid trees down to the east.

It's human nature to relish a good disaster—how else to explain the attraction of the Weather Channel?—and so it pained me even to consider the possibility that the hurricane hadn't been such a big deal here. Which is why Cogbill and I were looking at big trees and stepping down into enormous pits. He was intrigued by what he saw in our woods, but he wasn't ready to jump to any conclusions. "Get the data," Cogbill said. "And then we can see if something different happened here."

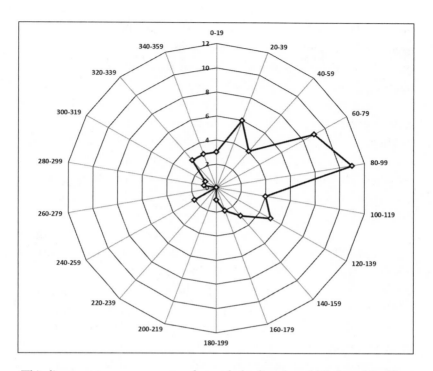

This diagram represents a compass face with the direction of fall plotted for fifty-eight trees on the author's woodland. Increased distance from the center indicates a higher number of trees that fell in that direction. Thirty-nine trees fell in a range from 20 to 159 degrees, from north to southeast. Only nine trees fell to the west.
Stephen Long

The data he wanted to see included a large sampling of the directions of the pits and mounds. The next day, I returned to the hill with my compass and notebook, recording compass bearings for fifty-eight of the most prominent pits and mounds. Some of them could hide a Volkswagen. At each stop, I also took a compass bearing for the aspect, the direction the hill faces. I also recorded the elevation at each pit and mound as I worked my way up the hill.

The data analysis confirmed what the eyeballs had said. The hill faces east and slightly south of east, with a median aspect of 115 degrees. The trees fell in a wide range of directions but with a preponderance of falls to the east; the median tree fall direction was 106 degrees. Emery Boose's HURRECON model predicted Corinth would have sustained winds of 78 miles per hour and gusts of 117 miles per hour from a direction of 104 degrees. Trees tipped over by those winds would be expected to fall at 284 degrees, almost the exact opposite of what happened to these hardwoods. In scientific parlance, the direction of fall was more closely correlated to aspect than to wind direction. The steeper the slope, the greater the tendency for trees to fall downhill. Of the thirty-six trees growing on a slope 20 percent or steeper, thirty-four of them fell downhill. Only five trees fell in the predicted north to northwest direction, four of which had stood less than sixty feet below the top of the ridge. There, as the slope diminished at the hilltop, gravity no longer opposed the wind direction, and the trees fell as expected. In addition, hurricane winds are stronger higher on a hilltop because friction has less chance to slow them.

I wanted to see whether our hillside was a freakish anomaly, so I visited Harry Brainerd's sugarbush, whose demise in 1938 he had so poignantly described to me. One of its slopes pitched to the east, just like ours, and on it twelve of fifteen trees fell downhill instead of to the northwest. Atop the hill, they fell to the north, as ours had. Emboldened, I visited Hubbard Brook, where I confirmed that many of the trees on their southeast-facing slopes had fallen downhill.

Any logger will tell you how difficult it is to fell a tree uphill. Gravity is working against you, and not only in the obvious way. If you examine crowns of trees growing on hillsides, you'll see that each one's mass is much greater on the side facing away from the slope. That's because it's not a level playing field. Each successive tree higher on the hill has an advantage over the one directly below. And each is in turn compromised by the one above. While a tree's

These trees are growing on a 30 percent grade that slopes from the left side of the photo down to the right. Their crowns are heavily weighted to the open downhill side. Stephen Long

backside is abraded by the tree above it, it develops a disproportionate amount of its crown on the open side. The mass of limbs and branches reaching for the light adds considerably more weight to the downhill side. That compounds the tree's inclination to fall downslope.

I was convinced that gravity is stronger than hurricane winds, and I had some data to back up my supposition. But I knew that white pine had fallen uphill on east-facing slopes that were every bit as steep as ours. I'd seen that clearly on one of the tracts at Harvard

Forest, known as Slab City, where the tip-up stumps clearly pointed uphill. I needed one more piece of evidence to show that hardwoods in northern New England had behaved differently.

The pits and mounds pointed downhill, but I hadn't proven *when* they had fallen downhill. The only way to know for certain was to determine how old the mounds were. That might seem like a daunting task, but it's quite straightforward. You can ascertain the age of a mound by determining the age of a tree growing on it. The disturbed soil of an exposed rootball provides an advantageous seedbed, so a tree growing on it, particularly one with prop roots, will be younger—with any luck, only slightly younger—than the mound. If a tree's rings show its origins to have been earlier than 1938, the mound predates the hurricane. I searched for paper birch or yellow birch because these species find mineral soil an optimal seedbed, but after a long search, I felt fortunate to find two mounds that had sugar maples growing directly on top. The dearth of newly seeded trees on mounds shows that in the posthurricane competition, most of the trees to reach the canopy started as advance regeneration between the pits and mounds.

I extracted cores with the increment borer, drilling into the tree as close to the ground as I could to reach the tree's first growth. I sanded until the samples were polished like pieces of fine furniture. I have to admit, my heart was racing when I brought the first core into focus on the microscope. I marked each decade with a pencil dot and went backward: 2010, 2000, 1990 . . . The first tree was from 1943, the second from 1942. Bingo. These maples had seeded onto mounds formed a few years earlier, when the trees blew down in 1938.

Still, I knew that I needed to replicate these results in another forest, so I went back to Hubbard Brook and was given permission to core some trees. I found four trees on downhill mounds, two white ashes and two yellow birches. I went through the same process, and the white ash trees turned out to have been seedlings in 1956 and 1960, which showed only that the mounds they grew on were older than that—not bad, but not at all conclusive.

The yellow birches were better bets, but the first one was a bust because its core had a small section of rot where the rings didn't show up. I had no way of knowing how many years were eaten up by that gap. But I hit pay dirt with the second yellow birch. This thirteen-inch-diameter tree was a seedling in 1942. It was standing on the mound of a tree that had fallen to the southeast (142 degrees). This birch and the two maples on my hillside showed that the mounds that provided their seedbed had been formed just before 1942. No question, they fell during the hurricane.

Charlie Cogbill helped paint a picture of what happened. Tall trees swayed as the sustained winds pushed them toward the hill and then let up. Rigid, older trunks flexed very little, transferring all the torque to the roots. Crowns bashed against one another as gusts rocked them even harder. At the height of the storm, gusts exceeding 100 miles per hour kept the trees swaying back and forth. Roots on both the downhill and uphill sides were alternately compressed, then tensed. "It was the tree rebounding back in the downhill direction that stretched the roots to the breaking point," Cogbill said. "The torque from the rebound was enough to break the roots, and they failed under tension on the uphill side." And down the trees went, some like dominoes, others on their own. And some, like Andrew Jackson, held their ground.

How extensive was this phenomenon of hardwoods falling downhill rather than downwind? And how steep did the slope need to be for it to happen? More data need to be collected to know for sure.

In my discussions with Charlie Cogbill and David Foster on the mechanics of the root failure, each of them independently questioned one of the truisms handed down over the years. Many accounts of the hurricane mention the preceding days of torrential

Damage to a New Hampshire farm woodlot between Salisbury and Franklin. The roots of the uprooted trees have clearly broken, rather than pulling free from the saturated ground. United States Forest Service

rain and draw the conclusion that the soggy ground made it easier for the trees to uproot. "While that makes intuitive sense," Foster said, "there is no evidence that the rain had any impact on uprooting. Soil scientists who have thought about this point out that in some soils, like sandy soils, adding moisture actually increases cohesion and should make trees more windfirm."

If it had been the ground that failed, the result would have been mudslides, Cogbill told me. "The roots didn't slip out of the supersaturated soil, they broke off. And they tended to break off at the point where the root tapered down to the size of your wrist."

Photos of fallen trees confirm this. From Connecticut to New Hampshire, from Rhode Island to Vermont, trees uprooted because their roots broke. The trunks acted as giant levers as the wind rocked them back and forth on a pivot point beneath the trunk. When the roots could no longer withstand the stretching or the compression, they snapped. Rain or no rain, the trees would have fallen.

TEN

The Wind Next Time

Count on it. Someday, another hurricane of Thirty-Eight's magnitude will hit New England.

Even though it has happened only three times in five centuries—in 1635, 1815, and 1938—meteorologists and ecologists view New England hurricanes as normal but rare. Normal means another corker can arrive any day now. Rare suggests we needn't bother spending every August through November prowling the widow's walk with an eye out to sea, because statistically it could just as easily hold off until the next century. That level of uncertainty makes planning for it impossible.

Let's suppose for a moment that it will come next hurricane season. What can we expect from the storm surge, the flooding, and the wind? And how will we respond to it?

Most significant, we won't be caught off guard as people were in 1938, so the death toll probably won't approach the six hundred lives lost then. Forecasters will let us know well in advance that the next one is on its way. Satellites provide such timely images that no tropical storm will ever again make a run for it and slam into Long Island and New England without its potential destruction being trumpeted through radio, TV, newspapers, and digital media, including tweets, emails, text warnings, and banner announcements from weather websites such as Intellicast and Accuweather. A greater danger than surprise is saturation, with the little boy crying hurricane so many times that nobody heeds his call.

Forecasters have become increasingly accurate in predicting a storm's track as it nears the coast, but their weather models are not as adept at gauging how damaging it will be once it makes landfall. Even when the beach umbrellas are already sailing and barometers plummeting, forecasters don't know how much force a storm will retain and for how long.

We do know what it will take for another hurricane to reach Petersham, Rindge, Hubbard Brook, and Corinth. It will require approximately the same characteristics as the previous three, which meteorologist Lourdes Aviles described as "fast, northward-moving storms that are still holding on to significant intensity in their core while transitioning into extratropical storms." An unusual configuration of high and low pressure systems will provide a hurricane path that approximates that of Thirty-Eight's, though perhaps without its uncanny westward pull. The next big one will track northward, cross Long Island and Long Island Sound, and make a second landfall in Connecticut or Rhode Island. Other paths always cause less damage. For instance, a storm that makes landfall on the mainland anywhere south of Long Island will lose its momentum, and its damage to New England will be muted. A storm that recurves to the east and out into the Atlantic, even if it brushes Cape Cod and the adjacent islands, will present itself with downpours but not much wind inland in New England.

The entry point of the next big one could be anywhere along the length of Long Island. For it to penetrate all the way into New

TABLE 3

POPULATION OF NEW ENGLAND STATES AND TWO OUTER COUNTIES OF
LONG ISLAND IN 1940 AND 2010

New England	1940	2010	1940 rural	2010 rural
Massachusetts	4,316,721	6,547,629	457,245	525,640
Connecticut	1,709,242	3,574,097	551,080	429,155
Rhode Island	713,346	1,052,567	59,963	97,524
Maine	847,226	1,328,361	504,169	814,819
New Hampshire	491,524	1,316,470	208,299	522,598
Vermont	359,231	625,741	235,992	382,356
Total	8,437,290	14,444,865	2,016,748	2,772,092
Long Island				
Nassau County	406,748	1,339,532		
Suffolk County	197,355	1,493,350		
Total	604,103	2,832,882		
Combined total	9,041,393	17,277,747		

Note: Nassau and Suffolk counties were not delineated into rural populations in 1940 or 2010, though in 1940 both had substantial rural populations.
Source: United States Census Bureau.

Hampshire and Vermont, it will need to maintain a forward speed of at least 40 miles per hour. Otherwise, friction and the loss of the ocean's warm air will sap any hurricane's strength. Aviles has shown that the hurricanes of 1635, 1815, and 1938 all sped forward with at least that speed.

The storm itself will be similar to those three major hurricanes, but the landscape it blows through is now vastly different. First, many more of us are living in the region. The population of the six New England states plus the two outer counties of Long Island (Nassau and Suffolk) was just over 9 million in 1940. Today it is 17.3 million, almost double. The single most vulnerable place, Long Island, today has nearly *five* times as many people living there. Next in line for the mayhem, Connecticut's population has doubled.

The property damage will be astronomical. Thirty-Eight has been deemed New England's most devastating weather event as calculated by the cost to replace everything that was destroyed. In 2008, a company called Risk Management Solutions, which studies risk for large insurance companies, published a report on the expected damage from a replay of Thirty-Eight. It pegged Thirty-Eight's losses at approximately $300 million, which when adjusted for inflation would be around $5 billion in current dollars. It makes a further adjustment to the potential toll of a repeat performance, taking into consideration the increased population of the region and the increased average wealth of the populace to arrive at an estimate of potential damage ranging from $37.3 billion to $39.2 billion. The National Weather Service arrives at a similar estimate for a replay of Thirty-Eight, pegging potential losses at $41 billion.

When the next big one comes, Long Island, Connecticut, and Rhode Island will once again absorb the full force of the storm, suffering the surge, the flood, and the wind. The population growth has brought with it an increase in houses, commercial buildings, vehicles, roads, and bridges. Any of these along the coast will be vulnerable to a surge comparable to Thirty-Eight's, which brought sixteen feet of water to downtown Providence. Despite the lessons of Thirty-Eight and other lesser storms, people continue to build houses as close to the beach as they're allowed. Beachfront communities disappeared in 1938 when the tremendous surge knocked houses from their foundations and sucked them back into the ocean. Many people who survived did so by clinging to floating roofs and walls as they were swept out to sea. Given the penchant for shore-front development, there's no reason to suppose that we won't see damage like that happen again.

If heavy rains precede and accompany the hurricane, river flooding will rival the storm surge in its capacity for chaos. As Irene and

Thirty-Eight both showed, a soaked sponge doesn't sop up any water. Once the soil is saturated, all of the rain will flow over the ground, rather than soak into it. Across New England, there are more than a hundred flood-control structures, many of them built since 1938. All of these dikes, floodwalls, earthen dams, and reservoirs are designed to contain the rising water and stop it from rushing downstream. Many of these easily overlooked facets of the landscape attract attention only when they do their job and fill up with water. They have helped considerably in controlling smaller floods, but when up against a hundred-year flood like Irene, they didn't control as much as people would have liked. In Irene, propane tanks and Subarus tumbled end-over-end down rivers that you might otherwise wade across without difficulty. Raging rivers took out roads and bridges in spectacular displays of nature's might. Even small streams high up in the watersheds gouged out new or deeper watercourses. Count on serious flooding once again.

Then, of course, there will be devastating wind. Much of the population increases have been in the urban and suburban areas along the coast, but 750,000 more people live in rural New England now than in 1940. The rural population boom hasn't resulted from a fecund native population. Instead, the gain has come from a migration of people leaving cities and suburbs for a less stressful country life. Accustomed to long commutes, the newly rural travel willingly for work, or increasingly they telecommute. Businesses that formerly could function only with the handy infrastructure of cities have taken root in home offices at the end of gravel roads, the beneficiaries of internet connections. The net result is that people are living in remote places that haven't been inhabited since agriculture ruled the land in the nineteenth century.

Consequently, more miles of power lines and telephone lines can be snapped by fallen trees, despite the best efforts of right-of-way crews to minimize the risk. Today's communications network seems powerful and impressive, but it will keep everyone informed only until the power goes out. That could happen quickly and extensively.

Devices like uninterruptable power sources solve the problem for short-term outages, and generators can keep the power on to freezers, water pumps, and other appliances. The communications network has fewer fallbacks. Wireless devices rely on a signal being sent from a router, which requires power. Don't take any comfort in your smart phone as a fallback. In the first place, many hilly towns in the New England states lack cell coverage. And cell towers need power, too. If one of those faux pine towers loses power or is toppled by high winds, phones that rely on it for a signal won't work. It's that simple. Land line service will also disappear if fallen trees bring down phone lines. Any portable phone that requires electricity will be out of service.

We might be surprised how much we have in common with Harry Brainerd in 1938, who watched his family's radio aerial plummet from the barn roof, severing radio contact with the outside world. We can expect that most people—city or country—will be without telephone service or electricity for days, perhaps a week or more. Battery-operated radios might in fact be the best bet for keeping abreast of conditions beyond your yard.

People are more mobile today, with millions of people driving every day. That means much more traffic will be disrupted. The interstate highways have all been built since the 1950s. Engineers have designed them with wide open corridors, and most of the miles have been constructed with a wide right-of-way excluding trees, so toppled trees won't affect them. Still, there are plenty of pinch points where trees will probably block the interstate. A bigger problem will be reaching or escaping the interstate because the roads leading to them will be just as vulnerable to falling trees as they were in 1938. Many people could be stranded temporarily, and it's worth noting that temporarily simply means not permanently. Remember the Georgia ice storm that stranded people on frozen highways in Atlanta for twenty-four hours? Or the people trying to leave New Orleans during Katrina? Congestion on that scale can happen at many constricted points in a region with seventeen million people.

The next hurricane to roar across Long Island and into New England will encounter a transformed forest. The amount of land covered by forest has increased in northern New England since 1938, most dramatically in Vermont, which caught up with its neighboring states in abandoning farmland, resulting in today's 80 percent forest cover. All told, 75 to 80 percent of New England is covered with forests.

Trees continue to mature as growth exceeds the harvest. Our forests have doubled in volume since the 1950s, which means that much of the forest is full of trees large enough to blow down. That New England's forests have matured as much as they have results largely from a different kind of forest owner, one who doesn't count so heavily on the forest to provide periodic income. The most significant change in rural demographics is that most forest owners today are accidental owners. They didn't intend to own a forest— when they bought or inherited the house in the country, it came surrounded by woods. Parcel sizes have shrunk, so many people own a woodland not considered large enough to manage. They harvest their woods with a light touch, or not at all. Corporate owners provide the notable exception. More averse to risk, they are much less likely to grow trees to biological maturity. But corporate holdings are a minority within the footprint of Thirty-Eight and the presumed footprint of the next one. In parcels averaging twenty acres or so, hundreds of thousands of families own twice as much forestland as do investors. In 1938, thirty thousand forest owners suffered losses from the wind. Next time, we will multiply that number by as much as five because so many more people own a piece of woods.

If you glance up from this book and take a look out your window, chances are you can see some trees. Chances are equally good that you can see some tall trees, vulnerable trees, trees that could very well be knocked down by 100 mile per hour winds. Think of the landscape of your daily life, your neighborhood, your route to work, your favorite places, and consider how vulnerable all of this would be to high wind. Picture the worst wind you've ever experienced and then double the wind speed.

In 1938, the places that escaped wind damage did so because the trees they had were few or too young to blow down, or they were protected by topography. The latter sites will be sheltered once again, but the exposed south- and east-facing slopes will again be in the crosshairs, along with the flat and gently rolling topography. The unprotected sites today are much more likely to have mature trees, so the devastation will be much worse. In 1938, all of that havoc in the woods played out on 600,000 acres. This time around, double that acreage, triple it, perhaps even multiply it by ten. Six million blown-down acres within a footprint of 15 million acres would not be out of the question.

What will we do in the face of such a catastrophe? We have a recent example of how this worked out in a different part of the country. When Hurricane Katrina hit, all eyes focused on New Orleans and the devastation to a grand old city and its uncommonly vulnerable citizens. Unless you lived in southeastern Louisiana, you might not have known that Katrina destroyed 3 billion board feet of timber, 60 percent of it yellow pine and other softwoods. Hurricane Rita, which hit southwestern Louisiana a month later, knocked down an additional 1.6 billion board feet. These tallies are for Louisiana only and don't include the substantial loss in neighboring states. Combined, those two hurricanes blew down nearly twice what went down in 1938.

Much of the affected forestland was owned by individuals in relatively small parcels. As elsewhere, including the Northeast, the logging infrastructure in Louisiana had diminished greatly. Before the hurricane, more of the timber was being harvested by fewer loggers, most of them using mechanized equipment. The loggers tended to work on the lands of the larger corporate owners, who had long-standing supply agreements with mills. The smaller operators, a logger or two with a chainsaw and a skidder, have been disappearing slowly from the woods.

That pattern held in the hurricane cleanup. Fewer loggers and more landowners with smaller holdings meant that there weren't enough loggers to salvage the wood. Loggers came in from northern Louisiana and neighboring states to help clean up the mess and make a buck. The Forest Service's fear in 1938—and one of its prime motivations for stepping in—came true with Katrina because many landowners had to either pay to have the work done or give the stumpage away.

While New England and Louisiana may seem to be worlds apart, we can expect a similar outcome because of striking similarities in the rural forest economy in the two regions. Given the steady erosion of New England's timber-based economy, the logging infrastructure isn't large enough to salvage a catastrophic blowdown. Mechanized harvesters can process prodigious amounts of wood, but they can't be in more than one place at a time. If hundreds of thousands of forest owners experience blowdowns, there simply aren't enough loggers to do anything about it.

Some of the cable skidders that sit rusting behind garages could be brought back to life. Similar to what happened in 1938, an amateur workforce could spring into action, a small army of weekend warriors armed this time with Husqvarna and Stihl chainsaws. New businesses will spring up offering to clean up the mess for landowners. Will they pay for the logs or will they charge the landowner for their services? The local market will tell.

Human nature tells us to fix something that has gone wrong. When Irene's floods washed out sections of more than two thousand roads in Vermont, Governor Peter Shumlin allowed emergency workers to drive excavators into streams to mine gravel and rearrange streambeds. Convoys of dump trucks brought in rock to armor streambanks with rip-rap. All of this was done under emergency powers in an attempt to bring the situation back to "normal." It reopened roads but in the process rolled back years of work by the state to allow rivers access to their floodplains in order to minimize the ill effects of flooding. Almost overnight, miscreant rivers were assigned new channels and told to stay there, a practice that has never worked out well.

New Englanders will be pulled by the same impulse. We'll feel a sense of responsibility to somehow right the wrong when the next Thirty-Eight rearranges our woods. Accidental owners might be so bothered by the tangled mess of the blowdown that they'll gladly pay for the cleanup.

Countering that is an increasingly insistent drumbeat that tells us that disturbance is natural—even disturbance of this magnitude—and that a blown-down forest will heal on its own. We know that without human intervention, the forest will retain ecological control over the site even though all of the biomass is suddenly within ten feet of the forest floor. It will govern the new normal.

Because people own these blown-down forests, economics cannot be ignored. Those not in a financial position to let nature take its course will doubtless choose to recoup as much of the loss as they can. Others who do a cost-benefit analysis of the situation, taking into consideration the increased cost and danger of salvage logging and the reduced stumpage value caused by a glut of wood, might see that the possible income from the salvage is just not worth it.

The government will not step in as it did in 1938. Instead, forestry departments will caution landowners not to act rashly. One in-

A tornado touched down in central Massachusetts in June 2011. Brimfield State Forest was one of the areas hit. Except for fifty-foot-wide firebreaks along roads and boundaries, none of the downed timber was salvaged. Note the circular pattern of the windthrow. William N. Hill, Massachusetts Bureau of Forestry

dicator of shifting governmental attitudes is the Massachusetts Bureau of Forestry's response to a tornado that blew through Brimfield State Forest in June 2011. Its foresters chose not to salvage the 940 acres blown down. Instead, it simply created firebreaks along roads and boundaries to ensure that any fire wouldn't spread. It designated the entire 3,523-acre forest a reserve and set up study plots to monitor the regrowth of the vegetation and the changing songbird populations in the wake of the disturbance. What a far cry from the hysteria whipped up in 1938.

Whether people salvage or not could have huge implications for the residual forest. Loss of forest cover means that less water penetrates the earth's surface. Increased water flow over large areas stripped of vegetation can erode land downhill. Further, salvage logging can open up the forest to an abrupt rather than gradual change in species composition. Poised on the edge of many forests are nonnative invasive shrubs and trees whose populations could explode following a salvage operation that bathes them in sunlight and scrapes up a welcoming seedbed.

In 1938, the concept of invasive species did not exist, and it wouldn't gain traction for another four decades. In fact, until the 1970s, fish and game departments across the nation were actively promoting nonnative shrubs as beneficial for wildlife. They handed landowners ready-to-plant autumn olive and Russian olive, bittersweet, buckthorn, several species of honeysuckle, Japanese barberry, and the most impossible of them all, multiflora rose, which is so impenetrable it was touted as the natural fence.

All of these species thrive in openings but don't readily spread into the understory beneath a full canopy. Once established, however, they can survive for many years after trees overtop them. In any place with a canopy opening near a seed source, buckthorn, honeysuckle, and the others will take hold. On the other hand, a forest opened up by a hurricane but left unsalvaged will not necessarily welcome new species because the existing vegetation continues to control the territory. But when a salvage operation follows, it carries the potential for the wholesale spread of these species. An attempt to return things to normal could instead open the door to a hostile takeover.

■

As we face our changing climate, it's hard not to be alarmed by weather events that seem freakish—too many subzero nights, too many October and April snowstorms, too much rain in a sudden downburst, too many presidential declarations of disaster areas. Hyperaware of extreme weather, we both relish and fear a good storm.

Still, we cannot be prepared for what the next Thirty-Eight will bring.

Once the storm turns northward, and the singular combination of weather fronts steers the storm right at the heart of New England, we will all face a continuous explosion. We are powerless to stop it and powerless to diminish its incredible destructive force. It will rip roofs from buildings; it will topple trees onto our houses, roads, and power lines; it will turn every stream into a torrent. And as it's ripping our world apart, it will relentlessly scream and roar, searing into our consciousness a sound that none of us will ever forget.

On that day, nature will make mankind feel puny. As the Taino knew, *Huracan* is an angry goddess, and her punishment can be severe.

Bibliography

Allen, Everett S. *A Wind to Shake the World: The Story of the 1938 Hurricane.* Beverly, Mass.: Commonwealth, 1976.

Amos, E. G. "Final Report on the New England Forest Emergency Fire Hazard Reduction Project." *Journal of Forestry* 39 (1941): 749–52.

Anderson, Dave. "Fire on the Mountain." *Forest Journal,* July 6, 2008.

Aviles, Lourdes. *Taken by Storm, 1938: A Social and Meteorological History of the Great New England Hurricane.* Boston: American Meteorological Society, 2013.

Barker, Rocky. *Scorched Earth: How the Fires of Yellowstone Changed America.* Washington, D.C.: Island, 2013.

Barker Plotkin, Audrey, David Foster, Joel Carlson, and Alison Magill. "Survivors, Not Invaders: Control Forest Development Following Simulated Hurricane." *Ecology* 94 (2013): 414–23.

Barlow, Virginia. Under the Microscope (series). *Northern Woodlands,* Summer 1994–Summer 2011.

Barraclough, Solon L. "Forest Land Ownership in New England: With Special Reference to Forest Holdings of Less than Five Thousand Acres." Ph.D. thesis, Harvard University, 1949.

"Barre Was Battered Whole Black Night by Terrific Wind." *Barre Daily Times,* September 24, 1938.

Bibliography

Barron, Hal. *Those Who Stayed Behind: Rural Society in Nineteenth-Century New England.* Cambridge: Cambridge University Press, 1984.

Barton, Andrew M., Alan S. White, and Charles V. Cogbill. *The Changing Nature of the Maine Woods.* Hanover, N.H.: University Press of New England, 2012.

Bell, Michael. "Did New England Go Downhill?" *Geographical Review* 79, no. 4 (1989): 450–66.

Black, John Donald. *The Rural Economy of New England: A Regional Study.* Cambridge: Harvard University Press, 1950.

Boose, Emery R., Kristen E. Chamberlin, and David R. Foster. "Landscape and Regional Impacts of Hurricanes in New England. *Ecological Monographs* 7 (2001): 27–48.

Boose, Emery R., David R. Foster, and Marcheterre Fluet. "Hurricane Impacts to Tropical and Temperate Forest Landscapes." *Ecological Monographs* 64 (1994): 369–400.

Bormann, F. Herbert, and M. F. Buell. "Old-Age Stand of Hemlock–Northern Hardwood Forest in Central Vermont." *Bulletin of the Torrey Botanical Club* 91 (1964): 451–65.

Bormann, F. Herbert, and Gene E. Likens. "Catastrophic Disturbance and the Steady State in Northern Hardwood Forests." *American Scientist* 67 (1979): 660–69.

———. *Pattern and Process in a Forested Ecosystem.* New York: Springer Verlag, 1994.

Brake, R. W., and H. A. Post. "Natural Restocking of Hurricane-Damaged 'Old Field' White Pine Areas in North Central Massachusetts." Master's thesis, Harvard University, 1941.

Brooks, Charles F. "Hurricanes into New England: Meteorology of the Storm of September 21, 1938." *Geography Review* 29 (1938): 119–27.

Burns, Cherie. *The Great Hurricane, 1938.* New York: Grove, 2005.

Butler, Brett J. *Family Forest Owners of the United States: A Technical Document Supporting the Forest Service 2010 RPA Assessment.* Newton Square, Pa.: USDA Forest Service, Northern Research Station, 2006.

Clark, Blair. "New Disaster of Fire, Coming from Fallen Wood, Predicted: Director of Harvard Forest Paints Ominous Picture of Devastation in Connecticut River Valley as Worst After-Effect of Disaster Wrought by Hurricane." *Harvard Crimson*, September 27, 1938.

Cline, A. C. *The Marketing of Lumber in New Hampshire 1925.* Cambridge: Harvard University Press, 1926.

Cogbill, Charles V., John Burk, and Glenn Motzkin. "The Forests of Presettlement New England, USA: Spatial and Compositional Patterns Based on Town Proprietor Surveys." *Journal of Biogeography* 29 (2002): 1279–1304.

Colby, Joseph G. Unpublished journal, 1924–42.

Cooper-Ellis, Sarah, David R. Foster, Gary Carlton, and Ann Lezberg. "Response of Forest Ecosystems to Catastrophic Wind: Evaluating Vegetation Recovery on an Experimental Hurricane." *Ecology* 80 (1999): 2683–96.

Cronon, William. *Changes in the Land: Indians, Colonists, and the Ecology of New England.* New York: Hill and Wang, 1983.

Downs, J. B. *The Wood-Using Industries of Massachusetts, 1926.* Harvard Forest Bulletin, no. 12. Petersham, Mass.: Harvard Forest, 1928.

Drury, William H., Jr. *Chance and Change: Ecology for Conservationists.* Berkeley: University of California Press, 1998.

Drury, William H., and Ian Nisbet. "Succession." *Journal of the Arnold Arboretum* 54 (1973): 331–68.

Egan, Timothy. *The Big Burn: Teddy Roosevelt and the Fire That Saved America.* New York: Houghton Mifflin Harcourt, 2009.

Egler, Frank. "Vegetation Science Concepts I: Initial Floristic Composition, a Factor in Old-Field Vegetation Development." *Vegetatio* 4 (1954): 412–17.

———. *The Way of Science: A Philosophy of Ecology for the Layman.* New York: Hafner, 1970.

Emmanuel, Kerry. *Divine Wind: The History and Science of Hurricanes.* New York: Oxford University Press, 2005.

Fisher, Richard T. "General Marketing Conditions in New Hampshire." *Harvard Forest Bulletin* 10 (1926): 7–12.

———. *The Harvard Forest: A Model Forest to Demonstrate the Practice of Forestry, an Experiment Station for Research in Forestry and Allied Problems, a Field Laboratory for Graduate Students.* Petersham, Mass.: Harvard Forest, 1929.

———. "New England Forests: Biological Factors." *New England's Prospect* 16 (1933): 213–23.

———. "Soil Changes and Silviculture of Harvard Forest." *Ecology* 9 (1928): 6–11.

Forest History Society. U.S. Forest Service History. www.foresthistory.org.

Foster, David R. "Species and Stand Response to Catastrophic Wind in Central New England, U.S.A." *Journal of Ecology* 76 (1988): 135–51.

Foster, David R., and John D. Aber. *Forests in Time: The Environmental Consequences of 1,000 Years of Change in New England.* New Haven: Yale University Press, 2004.

Foster, David R., John D. Aber, Jerry M. Melillo, Richard D. Bowden, and Fakhri A. Bazzaz. "Forest Response to Disturbance and Anthropogenic Stress." *BioScience* 47 (1997): 437–45.

Foster, David R., and Emery Boose. "Patterns of Forest Damage Resulting from Catastrophic Wind in Central New England, USA." *Journal of Ecology* 80 (1992): 79–99.

Foster, David R., Glenn Motzkin, and Benjamin Slater. "Land-Use History as Long-Term Broad-Scale Disturbance: Regional Forest Dynamics in Central New England." *Ecosystems* 1 (1998): 96–119.

Foster, David R., and John F. O'Keefe. *New England Forests through Time: Insights from the Harvard Forest Dioramas*. Cambridge: Harvard University Press, 2000.

Foster, David R., and David A. Orwig. "Preemptive and Salvage Harvesting of New England Forests: When Doing Nothing Is a Viable Alternative." *Conservation Biology* 20, no. 4 (2006): 959–70.

Freeman, Stan. "Deadly Storm Took New England by Surprise." *Republican Newsroom*, September 21, 2008. http://www.masslive.com/news/index .ssf/2008/09/deadly_storm_took_new_england.html.

Goodwin, Neil. *We Go as Captives: The Royalton Raid and the Shadow War on the Revolutionary Frontier*. Montpelier: Vermont Historical Society, 2010.

Gove, Bill. *J. E. Henry's Logging Railroads*. Littleton, N.H.: Bondcliff, 1998.

Greeley, W. B., Earle H. Clapp, Herbert A. Smith, Raphael Zon, W. N. Sparhawk, Ward Shepard, and J. Kittredge, Jr. "Timber: Mine or Crop?" *United States Department of Agriculture Yearbook, 1922*, 83–180. Washington, D.C.: Government Printing Office.

Grossi, Patricia. *The 1938 Great New England Hurricane: Looking to the Past to Understand Today's Risk*. Newark, Calif.: Risk Management Solutions, 2008.

Gullett, Georgiann. "Taking a Second Punch: Hurricane Rita." *Forests and People*, Third Quarter 2014, 10–12.

Harper, Roland M. "Changes in the Forest Area of New England in Three Centuries." *Journal of Forestry* 16: 442–52.

Hawes, Austin F. *Chestnut in Connecticut and the Improvement of the Woodlot*. Connecticut Experimental Forestry Station Bulletin 154, 1906.

———. "History of Forestry in Connecticut." Unpublished manuscript, Connecticut Forest and Park Association, 1957.

———. *Hurricane Damaged Forests: Still an Important State Asset*. Hartford: Connecticut State Forestry Department, 1939.

————. *The Present Conditions of Connecticut Forests, a Neglected Resource*. New Britain, Conn.: Record, 1933.

Henry, J. D., and J. M. A. Swan. "Reconstructing Forest History from Live and Dead Plant Material: An Approach to the Study of Forest Succession in Southwest New Hampshire." *Ecology* 55 (1974): 772–83.

Hibbs, David E. "Forty Years of Forest Succession in Central New England." *Ecology* 64 (1983): 1394–1401.

History of the Connecticut Valley in Massachusetts, vol. 2. Philadelphia, 1879.

Jones, Nard. "Man from the Hurricane." *American Forests*, February 1954, 16–18, 42.

Junger, Sebastian. *The Perfect Storm*. New York: Norton, 1997.

Kneipp, Leon F. "Land Planning and Acquisition, U.S. Forest Service." Transcript of an oral history conducted by Amelia R. Fry, Edith Mezirow, and Fern Ingersoll, 1964–65. Regional Oral History Office, Bancroft Library, University of California, Berkeley, 1976.

Kotok, E. I. and R. F. Hammatt. "Ferdinand Augustus Silcox." *Public Administration Review* 2 (1942): 240–53.

Landgraf, Walter. *Town of Madison Rockland Preserve Historic Charcoal Site on the Houston Nature Trail, 1700s–1930s*. Madison, Conn.

Larson, Erik. *Isaac's Storm: A Man, a Time, and the Deadliest Hurricane in History*. New York: Vintage, 1999.

Ludlum, David M. *Early American Hurricanes, 1492–1870*. Boston: American Meteorological Society, 1963.

Lutz, Harold J. *Trends and Silvicultural Significance of Upland Forest Successions in Southern New England*. Yale University School of Forestry Bulletin 22, 1928.

Maher, Neil M. *Nature's New Deal: The Civilian Conservation Corps and the Roots of the American Environmental Movement*. New York: Oxford University Press, 2008.

McCarthy, Joe. *Hurricane!* New York: American Heritage, 1969.

Meeks, Harold. *Vermont's Land and Resources*. Shelburne, Vt.: New England Press, 1986.

Merrens, Edward J., and David R. Peart. "Effects of Hurricane Damage on Individual Growth and Stand Structure in a Hardwood Forest in New Hampshire, USA." *Journal of Ecology* 80 (1992): 787–95.

Merrill, Perry H. *Roosevelt's Forest Army: A History of the Civilian Conservation Corps*. Montpelier, Vt.: Perry H. Merrill, 1981.

Miller, Char. *Gifford Pinchot and the Making of Modern Environmentalism*. Washington, D.C.: Island, 2013.

Minsinger, William Elliott, ed. *The 1938 Hurricane: An Historical and Pictorial Summary.* Milton, Mass.: Blue Hill Observatory, 1988.

Mooney, Charles M., H. L. Westover, and Frank Bennett. *Soil Survey of Merrimack County, New Hampshire.* Natural Resources Conservation Service, 1908.

National Hurricane Center website. www.nhc.noaa.gov.

National Oceanic and Atmospheric Administration (NOAA) website. www.noaa.gov.

National Weather Service website. www.weather.gov.

"N.E. Hurricane Kills 85." *Boston Daily Globe,* September 22, 1938.

NETSA (Northeastern Timber Salvage Administration). *Report of the U.S. Forest Service Programs Resulting from the New England Hurricane of September 21, 1938.* Boston: NETSA, 1943.

New England Forest Emergency Project and Northeastern Timber Salvage Administration, 1938–1943. National Archives at Boston. Waltham, Mass.

New Hampshire Disaster Emergency Committee. Report on Flood and Gale of September 1938. New Hampshire Disaster Emergency Committee, November 1938.

Oliver, Chadwick D., and Bruce C. Larson. *Forest Stand Dynamics.* New York: McGraw-Hill, 1990.

Oliver, Chadwick D., and Earl P. Stephens. "Reconstruction of a Mixed-Species Forest in Central New England." *Ecology* 58 (1977): 562–72.

Patric, James H. "River Flow Increases in Central New England after the Hurricane of 1938." *Journal of Forestry* 72 (1974): 21–25.

Peart, David R., Charles V. Cogbill, and Peter A. Palmiotto. "Effects of Logging History and Hurricane Damage on Canopy Structure in a Northern Hardwoods Forest." *Bulletin of the Torrey Botanical Club* 119 (1992): 29–38.

Peirce, Earl S. "Salvage Programs Following the 1938 Hurricane." Berkeley: University of California Bancroft Library Regional Oral History Office, 1968.

Perley, Sidney. *Historic Storms of New England.* Salem, Mass., 1891.

Pielke, R. A., J. Gratz, C. W. Landsea, D. Collins, M. A. Saunders, and R. Musulin. "Normalized Hurricane Damage in the United States, 1900–2005." *Natural Hazards Review* 9 (2008): 29–42.

Pierce, C. H. "The Meteorological History of the New England Hurricane of Sept. 21, 1938." *Monthly Weather Review* 67(1939): 237–85.

Powhatan Museum of Indigenous Arts and Culture website. www.powhatanmuseum.com.

Raup, Hugh M. *Forests in the Here and Now*. Missoula: University of Montana, 1981.

———. "Old Field Forests of Southeastern New England." *Journal of the Arnold Arboretum* 21 (1940): 266–73.

———. "The View from John Sanderson's Farm: A Perspective for the Use of the Land." *Forest History* 10 (1966): 2–11.

Rowlands, Willett. *Damage to Even-Aged Stands in Petersham, Massachusetts, by the 1938 Hurricane as Influenced by Stand Condition*. M.F. thesis, Harvard University, 1941.

Rudnicki, Mark. "Stand Structure Governs the Crown Collisions in Lodgepole Pine." *Canadian Journal of Forest Research* 33, no. 7 (2003): 1238–44.

Scotti, R. A. *Sudden Sea: The Great Hurricane of 1938*. New York: Back Bay, 2003.

Silcox, Ferdinand A. "A Challenge." *Service Bulletin* 21 (1937).

———. "Fire Prevention and Control on the National Forests. *Yearbook of Department of Agriculture for 1910*. Department of Agriculture, 1910.

———. "Guarding Democracy." *Service Bulletin* 23 (1939).

Smith, Sarah S. *They Sawed Up a Storm*. Portsmouth, N.H.: Peter E. Randall, 2010.

Spencer, Betty Goodwin. *The Big Blowup*. Caldwell, Idaho: Caxton, 1956.

Spurr, Stephen H. "Natural Restocking of Forests Following the 1938 Hurricane in Central New England. *Ecology* 37 (1956): 443–51.

Stephens, Earl P. "The Uprooting of Trees: A Forest Process. *Soil Science Society of America Proceedings* 20 (1956): 113–16.

Stockman, Ken. Letter to Putnam Blodgett. Private collection.

"Storm of August, 1900, in Burlington Worst on Record, Recalls Edgar Chiott." *Burlington Free Press*, September 23, 1938.

Tannehill, I. R. "Hurricane of September 16 to 22, 1938." *Monthly Weather Review* 66 (1938): 286–88.

Thoreau, Henry David. *The Succession of Forest Trees*. Thoreau Reader, http://thoreau.eserver.org.

Tompkins, Janet. "Southeast Louisiana Forests after Katrina." *Forests and People*, Third Quarter 2014, 4–7.

University of Vermont Extension Service. *Wood-Using Industries of Vermont*. Burlington: University of Vermont Extension Service, 1928.

Vallee, David R., and Michael R. Dion. *Southern New England Tropical Storms and Hurricanes, a Ninety-Eight Year Summary 1909–1997*. Taunton, Mass.: National Weather Service, 1998.

Vermont Department of Agriculture. *Vermont Maple Sugar and Syrup.* Bulletin of the Vermont Department of Agriculture no. 38, 1930.

Weishampel, John F., Jason B. Drake, Amanda Cooper, J. Bryan Blair, Michelle Hofton. "Forest Canopy Recovery from the 1938 Hurricane and Subsequent Salvage Damage Measured with Airborne LIDAR." *Remote Sensing of Environment* 109 (2007): 142–53.

Wessels, Tom. *Forest Forensics: A Field Guide to Reading the Forested Landscape.* Woodstock, Vt.: Countryman, 2010.

Whitney, Gordon. *From Coastal Wilderness to Fruited Plain: A History of Environmental Change in Temperate North America from 1500 to the Present.* Cambridge: Cambridge University Press, 1994.

Wilson, Harold Fisher. *The Hill Country of Northern New England: Its Social and Economic History, 1790–1930.* New York: Columbia University Press, 1936.

Wooster, Chuck. "Bedrock Values: The Underlying Differences between Vermont and New Hampshire," M.S. thesis, Dartmouth College, 1998.

Works Progress Administration/Federal Writers Project. *New England Hurricane: A Factual, Pictorial Record.* Works Progress Administration, 1938.

Zimmerman, Eliot. *A Historical Summary of State and Private Forestry in the U.S. Forest Service.* United States Forest Service, 1976.

Index

Page numbers in *italics* refer to illustrations or tables.